The Handbook
of Environmental Chemistry

Volume 4 Part A

Edited by O. Hutzinger

Air Pollution

With Contributions by
H. van Dop, P. Fabian, H. Güsten,
J. M. Hales, A. Wint

With 71 Figures

Springer-Verlag
Berlin Heidelberg GmbH

Prof. Dr. Otto Hutzinger

University of Bayreuth
Chair of Ecological Chemistry and Geochemistry
Postfach 3008, D-8580 Bayreuth
Federal Republic of Germany

ISBN 978-3-662-15205-8

Library of Congress Cataloging in Publication Data.
Main entry under title: The Handbook of environmental chemistry.
Includes bibliographies and indexes.
Contents: v. 1. The natural environment and the biogeochemical cycles / with contributions by P. Craig ... [et al.] –
[etc.] – v. 3. Anthropogenic compounds / with contributions by R. Anliker ... [et al.] – v. 4. Air pollution / with
contributions by H. van Dop ... [et al.]
1. Environmental chemistry – Collected works.
I. Hutzinger, O.
QD31.H335   574.5′222   [QHJ545.A1]   81-18272
ISBN 978-3-662-15205-8   ISBN 978-3-540-39222-4 (eBook)
DOI 10.1007/978-3-540-39222-4

2154/3140-543210

Preface to the Volume Series "Air Pollution"

Volume 4, Air Pollution, is a new series in the Handbook of Environmental Chemistry. One of the main features of the Handbook is its division into various subject matters. The series in volume 1 describes the Natural Environment and Biogeochemical Cycles. The second series of volumes deals with Reactions and Processes and the third lists and describes the most important Anthropogenic Compounds, that is, chemical compounds produced or released by man.

Air pollution problems could not be fitted easily into these categories. Atmospheric trace constituents are released from one or many sources, they may be transported long distances from these sources by atmospheric motion, they may be transformed chemically by reaction with other constituents in the air or ultraviolet light and, finally, they may be removed from the atmosphere by a variety of deposition mechanisms. Furthermore, during their residency in the atmosphere or after deposition certain trace constituents may have deleterious effects on the environment and human beings. This new series has been established to allow integrated treatment of these complex processes.

Volume 4, like the first three volumes is an open-ended series which will attempt, over the years, to present a complete description of various aspects of air pollution. It will be integrated with the other Handbook series through the choice of subject matters and with a system of cross referencing. I hope that this series like the earlier volumes will prove useful and find the approval of my colleagues.

Bayreuth, November 1985   Otto Hutzinger

General Preface

Environmental Chemistry is a relatively young science. Interest in this subject, however, is growing very rapidly and, although no agreement has been reached as yet about the exact content and limits of this interdisciplinary discipline, there appears to be increasing interest in seeing environmental topics which are based on chemistry embodied in this subject. One of the first objectives of Environmental Chemistry must be the study of the environment and of natural chemical processes which occur in the environment. A major purpose of this series on Environmental Chemistry, therefore, is to present a reasonably uniform view of various aspects of the chemistry of the environment and chemical reactions occurring in the environment.

The industrial activities of man have given a new dimension to Environmental Chemistry. We have now synthesized and described over five million chemical compounds and chemical industry produces about hundred and fifty million tons of synthetic chemicals annually. We ship billions of tons of oil per year and through mining operations and other geophysical modifications, large quantities of inorganic and organic materials are released from their natural deposits. Cities and metropolitan areas of up to 15 million inhabitants produce large quantities of waste in relatively small and confined areas. Much of the chemical products and waste products of modern society are released into the environment either during production, storage, transport, use or ultimate disposal. These released materials participate in natural cycles and reactions and frequently lead to interference and disturbance of natural systems.

Environmental Chemistry is concerned with *reactions in the environment*. It is about distribution and equilibria between environmental compartments. It is about reactions, pathways, thermodynamics and kinetics. An important purpose of this Handbook is to aid understanding of the basic distribution and chemical reaction processes which occur in the environment.

Laws regulating toxic substances in various countries are designed to assess and control risk of chemicals to man and his environment. Science can contribute in two areas to this assessment; firstly in the area to toxicology and secondly in the area of chemical exposure. The available concentration ("environmental exposure concentration") depends on the fate of chemical compounds in the environment and thus their distribution and reaction behaviour in the environment. One very important contribution of Environmental Chemistry to the above mentioned toxic substances laws is to develop laboratory test methods, or mathematical correlations and models that predict the environmental fate of new chemical compounds. The third purpose of this Handbook is to help in the basic understanding and development of such test methods and models.

The last explicit purpose of the Handbook is to present, in concise form, the most important properties relating to environmental chemistry and hazard assessment for the most important series of chemical compounds.

At the moment three volumes of the Handbook are planned. Volume 1 deals with the natural environment and the biogeochemical cycles therein, including some background information such as energetics and ecology. Volume 2 is concerned with reactions and processes in the environment and deals with physical factors such as transport and adsorption, and chemical, photochemical and biochemical reactions in the environment, as well as some aspects of pharmacokinetics and metabolism within organisms. Volume 3 deals with anthropogenic compounds, their chemical backgrounds, production methods and information about their use, their environmental behaviour, analytical methodology and some important aspects of their toxic effects. The material for volume 1, 2 and 3 was each more than could easily be fitted into a single volume, and for this reason, as well as for the purpose of rapid publication of available manuscripts, all three volumes are published as volume series (see Preface to Parts C and D of the Handbook). Publisher and editor hope to keep materials of the volumes one to three up to date and to extend coverage in the subject areas by publishing further parts in the future. Readers are encouraged to offer suggestions and advice as to future editions of "The Handbook of Environmental Chemistry".

Most chapters in the Handbook are written to a fairly advanced level and should be of interest to the graduate student and practising scientist. I also hope that the subject matter treated will be of interest to people outside chemistry and to scientists in industry as well as government and regulatory bodies. It would be very satisfying for me to see the books used as a basis for developing graduate courses on Environmental Chemistry.

Due to the breadth of the subject matter, it was not easy to edit this Handbook. Specialists had to be found in quite different areas of science who were willing to contribute a chapter within the prescribed schedule. It is with great satisfaction that I thank all authors for their understanding and for devoting their time to this effort. Special thanks are due to the Springer publishing house and finally I like to thank my family, students and colleagues for being so patient with me during several critical phases of preparation for the Handbook, and to some colleagues and the secretaries for technical help.

I consider it a privilege to see my chosen subject grow. My interest in Environmental Chemistry dates back to my early college days in Vienna. I received significant impulses during my postdoctoral period at the University of California and my interest slowly developed during my time with the National Research Council of Canada, before I could devote my full time to Environmental Chemistry in Amsterdam. I hope this Handbook may help deepen the interest of other scientists in this subject.

Otto Hutzinger

Contents

List of Contributors

Dr. H. van Dop
Royal Netherlands
Meteorological Institute
P.O. Box 201
NL-3730 AE De Bilt
The Netherlands

Prof. Dr. P. Fabian
Max-Planck-Institut für Aeronomie
D-3411 Katlenburg-Lindau
Federal Republic of Germany

Dr. H. Güsten
Kernforschungszentrum Karlsruhe
Institut für Radiochemie
D-7500 Karlsruhe
Federal Republic of Germany

Dr. J. M. Hales
Atmospheric Sciences Center
Earth Sciences Department
Battelle
Pacific Northwest Laboratories
P.O. Box 999
Richland, WA 99352, USA

Dr. A. Wint
Chemical Engineering Department
University of Nottingham
Nottingham NG7 2RD, England

Air Pollution in Perspective

A. Wint

Chemical Engineering Department, University of Nottingham
Nottingham, NG7 2RD, England

Summary

For many years man has been exposed to a variety of atmospheric pollutants. Until about 1960 the only effects causing strong concern occurred within a few kilometres of the sources of emission. Acceptable remedies included dispersion of effluent gases high into the atmosphere. In recent years it has become clear that many pollutants discharged in this way are being transported over large distances and are affecting many countries, for example in "acid rain". Similar global concerns include carbon dioxide and chlorofluorocarbon emissions, which may affect the balance of the atmosphere so as to cause climatic changes. With these problems there is still much uncertainty, especially regarding interactions between pollutants within the atmosphere and their combined effects on the environment. Compared with the local issues of the previous era, the scale and cost of each problem are very high. When a good understanding is reached, there may still be delays in reaching the necessary international consensus for effective action.

Introduction

Pollution of the atmosphere has often been a source of immediate worry to the people affected. An average adult breathes approximately 10 to 15 m^3 of air each day and has little choice whether or not to take the air in the vicinity. Throughout past ages, there has been an association between air pollution and the spread of disease, exemplified by the smells arising from insanitary conditions. In more recent times, concern about the effects of atmospheric pollution has widened to include damage to buildings and materials, crops, forests, animal species, freshwater ecosystems and even the stratosphere. Certainly much progress has been made in reducing the problems caused by many sources of pollution, especially those arising in industry and having most impact in a limited locality. However, in recent years, perceptions of the principal atmospheric pollution problems have changed. Many of the issues involve long-term effects on components of the biosphere or the atmosphere. Many are no longer simply local, but are on a regional or global scale [1].

A vast range of man's activities can discharge materials into the atmosphere which do not exist there normally. Many of these have no deleterious effects, at least on the basis of current knowledge, and can be excluded from this study. However, many other discharges must be included. An excellent definition of pollution has been put forward recently as "the introduction by man into the environment of substances or energy liable to cause hazards to human health, harm to living resources and ecological systems, damage to structures or amenity, or interference with legitimate uses of the environment" [2].

In the past, governments introduced systems to monitor and restrict emissions to atmosphere, in order to resolve conflicts of interest between the emitters of pollutants, often industrial firms, and the neighbouring population who were subject to the effects. These control systems were commonly aimed to achieve a balance between the costs imposed on the industrial processes and the resultant benefits to the community at large. The nature of the problems which have come to the forefront of attention in recent years makes such a balance much more difficult to attain. There are international issues, with countries holding widely differing viewpoints because of such factors as prevailing wind directions carrying airborne pollutants preferentially in one direction rather than the opposite, contrary opinions on the development of automobile engines in order to improve exhaust emissions and different interpretations of the role of law in such issues. Within nations too, there are aspects of such problems which raise strong political overtones, for example when choices have to be made between alternative fuels for generating electricity. While scientific research can provide information to enable some decisions to be made wholly objectively, others may have wide social and economic implications, full of controversy. It has been said that "pollution is but one thread in the tangled web of interactions that is modern society and one which cannot be neatly pulled out, leaving the remainder of the pattern undisturbed" [1].

The aim of this chapter is to provide a brief overview of current problems in air pollution control. As in many other fields, a study of how problems have arisen in the past and how they have been overcome may be helpful in deciding future courses of action. Because pollutants have become so numerous, the Royal

Society classification of airborne pollutants is described, with brief details of the more important substances under current discussion. A section of the chapter is devoted to a review of the effects on health of air pollution, especially possible links with the incidence of lung cancer. Finally data are presented which relate on the one hand to the estimated costs imposed on society (particularly on industrial processes) by means of pollution control measures and on the other hand to the estimated damage being caused by the current level of airborne pollution.

A Brief History of Air Pollution

Early Problems and Control Measures

This account concentrates almost entirely on experiences within the United Kingdom until about 1960, before broadening into a description of the worldwide problems of recent years. Up to this transition, both industrial production and domestic consumption were relatively small in size. Emissions to atmosphere were similarly modest in quantity so that air pollution was a problem that crossed international boundaries only rarely. Within this period, the United Kingdom serves as a good case for study, since it was well endowed with coal. This enabled it to be one of the first countries to enter the Industrial Revolution but was also a major contributory factor in leading it to experience some of the earliest local pollution problems.

Some pollution of the atmosphere occurs through natural causes, such as emissions from volcanoes, sulphurous springs, sea-spray and decaying organic matter in oceans, marshes and estuaries. However, with the possible exception of the hydrocarbons arising from the organic material, these are unlikely to have significant effects on man's activities, compared with man-made pollutants [3].

As early as the Middle Ages, the burning of coal in England was creating a smoke problem to people living in towns. In 1273 an ordinance prohibited the use of coal in London as being prejudicial to health and, in 1306, a royal proclamation forbade the use of sea-coal in furnaces. Approximately three hundred years later, there was a prohibition on the use of coal in London whilst Parliament was sitting. At about the same time, according to a Nottingham legend, Queen Elizabeth I decided to cut short her stay at Nottingham Castle, because of the unpleasant effects she was suffering from the smoke given off by the citizens' domestic fires.

Throughout the nineteenth century, a growing conflict developed between the emitters of smoke and nearby residents who were affected by it [4]. In that era, the emission of smoke was a source of pride to factory owners, as a sign of continuing industry and prosperity, and they did not recognise that smoke was a threat to health. However, steps were taken to reduce the immediate impact of the smoke on the local community by erecting large chimneys, themselves an additional cause for owners' pride.

In the same period, problems caused by emissions from metallurgical and chemical processes began to appear. Probably the most significant and noxious of such emissions came from a plant on Merseyside which James Muspratt had built in order to convert common salt into sodium sulphate:

$$2\,NaCl + H_2SO_4 \rightarrow Na_2SO_4 + 2\,HCl.$$

Muspratt's simple plan was to allow the hydrogen chloride by-product to escape from the furnace as a gas, then to pass directly into the atmosphere via a chimney. Naturally this very nasty effluent gas made a considerable impact on the neighbourhood and eventually Muspratt was taken to court, in 1838, the proceedings being described as follows. "Witness followed witness, testifying to painful physical sensations, tarnished door-knockers and fireplace fenders, dead poplars, elms and sycamores, damaged broccoli in Everton, and changes in land values resulting from the continued existence of the nuisance" [5]. The defence claimed that an adverse verdict would mean the extinction of the alkali trade in England. How many times have similar opposing arguments been put forward in other local pollution issues around the world in the years since then?

The result of the 1838 court case was that Muspratt was found guilty of causing a public nuisance. Nevertheless he was permitted to continue operating his process, as also were other alkali plants in this area and in Northumberland and Lanarkshire. Public discontent with the ensuing pollution simmered for a number of years, including landowners whose crops suffered severe damage. Eventually, one of these landowners provided the impetus for a compaign which resulted in the Alkali Act of 1863. Thus the Alkali Inspectorate was set up to control gaseous emissions of this type. It has been estimated that acid emissions from alkali plants in England and Wales were reduced from about 1,000 tons per week to 43 tons per week within the first year of the Inspectorate's formation [6]. Legislation since 1863 has widened the scope of the system to include many chemical and mineral processing plants and also electricity works [7]. Now, following a recent change of name, the Industrial Air Pollution Inspectorate controls the emissions to atmosphere from approximately 2000 industrial premises in England and Wales [8].

Throughout the nineteenth century and the first half of the twentieth century, progress in reducing atmospheric pollution from domestic sources and from small-scale industrial steam-raising plants was slow. Although various groups had been formed from time to time to promote smoke abatement, their opinions were often received with apathy. During short spells of still weather in various metropolitan areas, both in the United Kingdom and elsewhere, pollutant concentrations increased temporarily to such an extent that deaths from bronchial ailments had climbed significantly, e.g. in London (1873, 1880, 1891, and 1948), Glasgow (1909), Manchester and Salford (1930–31), the Meuse valley, Belgium (1930) and Donora, USA (1948) [4]. Then, in an inversion lasting four days in December, 1952, London became enveloped in a highly unpleasant "smog". At the time there was little reaction from the residents. However, this attitude changed abruptly in the following months after it had been shown that the smog had corresponded with a jump in the death rate of Greater London's population. Approximately 4000 additional deaths had occurred, mostly among the very young and the old.

The ensuing outcry led to the setting up of the Beaver Committee on Air Pollution and eventually to the Clean Air Act (1956). This, in turn, led to the estab-

Table 1. Deaths in greater London attributed to air pollution during comparable periods of atmospheric inversion in 1952, 1962, and 1972

Year	Max. mean daily concentrations in Central London ($\mu g/m^3$)		Estimated extra deaths in Greater London
	Smoke	SO_2	
1952	>6,000	3,500	4,000
1962	3,000	3,500	750
1972	200	1,200	nil

Reproduced from Royal Commissions on Environmental Pollution, 4th Report [9]

lishment of smokeless zones in cities and limits to grit, dust and smoke emissions from the combustion of fuels.

The ensuing outcry led to the setting up of the Beaver Committee on Air Pollution and eventually to the Clean Air Act (1956). This, in turn, led to the establishment of smokeless zones in cities and limits to grit, dust and smoke emissions from the combustion of fuels. Substantial improvements in urban air quality occurred in the United Kingdom in the years following that Act, as demonstrated by Table 1, which is reproduced from the Fourth Report of the Royal Commission on Environmental Pollution [9].

Table 1 shows estimates of the extra deaths in Greater London thought to have been directly attributable to air pollution during comparable periods of poor atmospheric dispersion due to temperature inversions, during the Decembers of 1952, 1962 and 1972. Clearly the concentrations of smoke had decreased sharply while SO_2 concentrations had declined less markedly. By 1972, the air quality had improved to such an extent that the temperature inversion had no noticeable effect on the normal death rate of the population.

In many other countries where the burning of coal had created risks to health, similar legislation was drawn up during this period and reductions in atmospheric pollution were obtained. Nevertheless new problems began to come into prominence.

Problems of Recent Years

From the nineteen-forties onwards, Los Angeles and some other metropolitan areas in the United States began to experience a "photochemical smog" during sunny days with poor atmospheric dispersion [10]. An increasing number of such occurrences have taken place in recent years, notably in the United States, Australia, Japan and Europe. Unlike the London smog, there has been no evidence to suggest a simultaneous increase in the death rate of the population affected. Nevertheless there may have been aggravation of existing chronic bronchial ailments [11] and certainly marked irritation of the eyes and throat, damage to vegetation and reduced visibility have occurred. Some time elapsed before the main cause was attributed to the interaction between hydrocarbons and nitric oxide in the presence of sunlight, yielding a complex mixture of pollutants including ozone, nitrogen dioxide and peroxyacetyl nitrate. It was realised that, in large conurbations, a considerable proportion of the primary pollutants, hydrocarbons and nitric oxide, was originating in automobile emissions, thus being distributed close to ground-level where they could create the maximum nuisance. In the United States, the problem has been tackled by equipping vehicles with catalytic converters, with the object of destroying unburnt hydrocarbons in the exhaust gases by oxidation. Within the European Economic Community, there is some controversy regarding the best course of action. Many countries favour adopting exhaust gas catalytic converters, while a body of opinion in the United Kingdom would prefer a "lean-burn" petrol engine, working with a high compression ratio, low ratio of fuel to air, aiming primarily to reduce emissions of nitrogen oxides.

Another complex problem which has caused deep concern in the industrialised parts of the world since the early nineteen-seventies is that of "acid rain". At that time it was noted that highly acidic rain and snow were being deposited in Scandinavia and parts of the United States, affecting water quality, fish life and forest productivity. Since then the reported ill-effects have spread to many other areas and have increased both in number and severity. There is evidence of large losses of fish from inland lakes [12], damage to considerable areas of forest [13], and also to crops, buildings and visibility. "Acid rain has become a focus of public concern, nationally and internationally" [14].

Even after several years' research effort, neither the formation of acid rain nor its effects on the environment are as yet fully understood. Unlike earlier pollution problems in history, the sources of the pollutants may be at large distances from the affected areas and our knowledge of transport processes through the atmosphere is still sketchy. Probably the chief primary causes of acid rain are sulphur dioxide and nitrogen oxide. During several days' residence in the atmosphere, these may be partly transformed into secondary pollutants such as sulphates and sulphuric acid, nitrates and nitric acid. Other pollutants, such as the oxidants implicated in the photochemical smogs, may also play important roles. Certainly it is clear that man is faced with a much more complex air pollution problem than in previous eras. The international implications are much more serious also. There is understandable resentment in countries subject to the worst effects, when a considerable proportion of the precursors to the problem may be emitted elsewhere. Removal of these precursors at source would not be easy nor cheap. Countries not suffering too badly are reluctant to spend large sums of money until a fuller understanding of the whole problem is arrived at.

Two other atmospheric pollution problems which have attracted attention in recent years are unusual because man may not have been seriously affected as yet. Nevertheless, if the more pessimistic predictions come true, the problems could become of crucial importance to the entire world. One of these problems centres on carbon dioxide and the "greenhouse effect", the other is concerned with the possible depletion of ozone in the stratosphere. Both problems will be outlined in the following section of this chapter.

Classification of Air Pollution Problems

For hundreds of years man has caused substances to be released into the atmosphere. However, the nature of these releases has changed in recent times owing to the increasing scale and variety of industrial processes, to the higher rates of consumption of energy and materials by many of the world's inhabitants and to the increasing complexity of the chemicals involved both in industrial and domestic activities. As described in the previous section, measures to control the impact of these emissions have been built up in individual countries mainly in terms of preventing all ill-effects in the locality of each source of emission. In general this has been achieved with reasonable success. Nevertheless there has been growing concern in recent years at our lack of understanding of the effects of some pollutants on a worldwide basis. In 1975, the Royal Society had set up a Study Group to undertake an intensive study of the long-range dispersal, absorption and chemistry of pollutants in the atmosphere and their effects on the environment. A report of their work was published in 1978 [15].

The Royal Society Study Group classified pollution issues as being global, regional (up to about 1,000 km from the source) or local (up to about 100 km from the source). This is a useful classification in serving to help clarify a very complex subject involving many substances. A minor drawback is that a few pollutants can be considered in more than one of these classes.

Outlines of a small number of specific examples will be used to illustrate each class of problem. For more detail, the reader is referred to other chapters in this series.

Global Air Pollution Problems

Carbon dioxide is usually present in the atmosphere in only dilute concentrations. However, the normal balance set up by nature is being gradually affected by the increasing quantities of carbon dioxide emitted during the burning of fossil fuels and also by deforestation processes, mainly in the tropics, in which vegetation is broken down either by natural decomposition or burning. The burning of fossil fuels has been estimated to have released 5.3×10^9 tonnes per annum of carbon dioxide into the world's atmosphere in 1980 [16]; deforestation, with less certainty, has been estimated to emit about 2×10^9 tonnes per annum [17]. However, these rates do not lead to an equal rate of accumulation in the atmosphere, since its content of carbon dioxide is taken up by green plants through photosynthesis and by the world's oceans and surface waters [18]. Nevertheless there is widespread agreement that, superimposed on normal seasonal fluctuations, the atmospheric concentration of carbon dioxide is undergoing a steady rise. For example, in Hawaii, there was an increase in the annual average concentration from 316 ppm in 1959 to 338 ppm in 1980 [19].

The effects of a rise in carbon dioxide levels are the subject of much conjecture. There is now a reasonable consensus that one consequence is a warming of the earth's surface and lower atmosphere by the so-called "greenhouse effect". This is due to the transparency of carbon dioxide to incoming short-wave solar radiation and its opacity to the longer-wave radiation back from the earth. However there is much uncertainty about the magnitude of the temperature increase, because of our inadequate knowledge of the influences of the oceans, clouds, particulates and other minor constituents of the atmosphere, such as ozone, chlorofluorocarbons, methane and nitrous oxide. Rainfall patterns could be affected also [20]. Substantial climatic changes would have a marked impact on many areas of the world, affecting agricultural yields, cattle and other animal species; perhaps even portions of polar ice would be melted, with the frightening possibility of raising the sea level appreciably. While the risks of short-term ill-effects are probably slight, there is an urgent need to improve predictive models of atmospheric behaviour, so that better understanding of all possible effects on climate can be achieved. Worthwhile reductions of carbon dioxide emissions could only be brought about by major changes in fuel and energy policies among the world's industrialised nations and such alterations would be almost impossible to achieve quickly.

Another potential global problem concerns ozone which is a minor constituent of the atmosphere. It occurs mainly in the stratosphere, at altitudes from 15 to 50 km above the Earth's surface and plays a key role, for the well-being of mankind, in filtering out much of the ultraviolet radiation approaching Earth from space. The concentration of ozone in the stratosphere is subject to considerable variation with latitude, longitude and altitude, but basically is governed by a dynamic equilibrium involving oxygen and a number of other gases present in trace quantities. Among the most significant of this group are chlorofluorohydrocarbons, which are used as aerosol spray propellants, refrigerants and foaming agents for plastics. When released, these compounds, which are noted for their high stability, tend to break down in the stratosphere, leading to the formation of chemical species which catalyse the destruction of ozone [21, 22].

The chief implications of a reduction in stratospheric ozone is that more ultraviolet radiation would reach Earth, perhaps affecting climates and significantly increasing rates of skin cancer. Since the danger was recognised, action has been taken in the United States and in the European Economic Community to effect a decrease in the use of chlorofluorocarbons as aerosol propellants. However there has been a small but significant increase in other applications and, more worryingly, production rates in other areas of the world have increased noticeably [23]. Clearly, in a global problem of this kind, effective action must be taken worldwide. Also monitoring and research programmes must include any materials which may affect the ozone concentration in a similar manner, for instance other halogenated organic compounds, methane and nitrogen oxides.

Regional Air Pollution Problems

Until recent years, almost all the effects of atmospheric pollutants were confined to areas within 100 km of the sources. Where emissions were from point sources, as for example from industrial plants, it was considered to be an acceptable arrangement to discharge at some height, via chimney stacks. Thus, using atmospheric dispersion, very high ground level concentrations near the sources of emission could be avoided.

Table 2. Emissions to atmosphere from the UK of some gaseous pollutants (million tons per year)

		1972	1973	1974	1975	1976	1977	1978	1979	1980	1981	1982
SO_2	Power Stations	2.87	3.02	2.78	2.82	2.69	2.74	2.81	3.10	2.87	2.71	2.60
	All sources	5.64	5.80	5.35	5.13	4.98	4.98	5.02	5.34	4.67	4.23	4.00
NO_2	Power Stations	0.73	0.81	0.72	0.76	0.79	0.79	0.81	0.88	0.85	0.82	0.77
	Vehicles[a]	0.42	0.45	0.44	0.47	0.45	0.46	0.48	0.49	0.49	0.48	0.49
	All sources	1.73	1.85	1.72	1.70	1.74	1.77	1.80	1.89	1.79	1.71	1.67
CO	All sources[a,b]	7.86	8.30	8.07	7.80	8.06	8.27	8.62	8.78	8.85	8.62	8.83
Total hydrocarbons (methane equivalent weight)		2.40	2.60	2.73	2.77	2.93	3.03	3.16	3.35	3.37	3.36	3.29

[a] These are the best estimates currently in a consistent format, but recent work at Warren Spring Laboratory suggests that vehicles emit somewhat more NO_x ($\leq 40\%$ of total) and and significantly less CO

[b] Principally vehicles

NO_x represented as equivalent NO_2

Reproduced from "Acid Rain" [14]

In effect, discharging through stacks served to spread pollutants through the atmosphere in more dilute concentrations. Over the last two decades however, as industrial output has grown, the total pollutant load on the atmosphere has risen to such an extent that harmful effects may be occurring at large distances from sources, and even across international boundaries. The foremost example of this type of problem is that of "acid rain", as mentioned in a previous section. The causes are complex and still not fully understood, but it is clear that a major contribution comes from sulphur dioxide. In northern Britain, as in Scandinavia, the average annual associations of acidity with sulphate and nitrate ions are about 70 per cent and 30 per cent respectively [1, 22, 23].

Large quantities of sulphur dioxide are produced by the combustion of fuels containing sulphur. Table 2 gives estimates of emissions of some of the principal gaseous pollutants discharged from within the United Kingdom, during the years 1972 to 1982 inclusive [14]. Although the rate of discharge of sulphur dioxide is high, its concentration among the other gaseous products of combustion is low and hence its removal is expensive. In the past it was usually accepted that these gases could be discharged to atmosphere through a stack without treatment for removing sulphur dioxide, provided that the stack was designed so as to protect the immediate surroundings from harmful concentrations.

Systematic measurements to aid the study of the spread of airborne pollutants have been in existence for a considerable period. In Europe a network of stations, the European atmospheric chemistry network, was first set up in 1955 to take samples of air and rain. This was reorganised in 1972, with ten member countries of the OECD participating in a programme to measure the long-range transport of air pollution. A further programme, the European Monitoring and Evaluation Programme, was started in 1977, with the inclusion of many of the Eastern European countries. At about the same time, a similar scheme was set up by the United States and Canada.

It is now thought that about one half of the sulphur dioxide emitted into the turbulent mixing layer of the atmosphere undergoes dry deposition by adsorption, impaction and sedimentation on vegetation, soil, buildings or water [14]. This is likely to occur within about 30 hours of emission, by which time it may have travelled up to 1,000 km. Much of the remainder comes to earth during wet deposition, associated with rain, snow or fog. This latter form is much more variable in nature and can lead to substantial deposits of high acidity in comparatively short times [1]. At present there is some uncertainty regarding the effects of these short-term episodes on soils and freshwaters compared with long-term rates of deposition.

Models have been developed which incorporate current understanding of the long-range transport of airborne pollutants. Predictions of annual deposition rates from these models are in reasonable agreement with measured rates. The models have been used to draw up a budget of sulphur emissions and depositions throughout Europe, demonstrating marked inequalities in the experiences of different countries. For example, one model indicates that in 1978–79, the United Kingdom received an average deposition rate of $4.2 \text{ g m}^{-2} \text{ y}^{-1}$ of sulphur, of which over 75% had been emitted from within its boundaries. By contrast, the Netherlands received $5.1 \text{ g m}^{-2} \text{ y}^{-1}$, of which less than 25% was of Dutch origin [14].

Several European countries have accepted a commitment to reduce sulphur dioxide emissions by 30% between 1980 and 1993, but the United Kingdom has been slower to agree. Table 2 shows that some of the most likely targets for effective action would be the electricity power stations, since they emit approximately 65% of the United Kingdom's total. A survey of the feasible control options has been made [24]. Halving the total emissions from power stations could be achieved by applying flue gas desulphurisation to the ten largest existing coal-fired stations, at a capital cost of $£1.5 \times 10^9$ (1982 prices). To pay for this, the price of electricity throughout the United Kingdom would have to rise by about 6% [1, 25].

Opposition to implementing a plan of this type in the United Kingdom is based on doubts as to whether a reduction in sulphur dioxide emissions would produce a proportional decrease in the harmful effects of "acid rain" on the environment. Particular points of uncertainty are:

(i)   the contribution of sulphur dioxide to damage of the environment compared with other atmospheric containments such as nitrogen oxides, hydrocarbons and ozone [12, 26, 17];
(ii)   the behaviour of chemical transformations within the atmosphere;
(iii)   the importance and unpredictable nature of short-term deposition of high activity;
(iv)   whether money spent on reducing emissions from new power stations, using pressurised fluidised bed technology, would be more effective than an equal investment on existing stations [28].

Further research is required to establish the best course of action, which must involve international consensus.

Local Air Pollution Problems

Substances which are limited in their impact to an area within 100 km of their point of emission form the vast majority of atmospheric pollutants. They can arise from a variety of sources and can comprise a wide range of chemical compounds in gaseous, liquid or solid form.

Until recent years, air pollution control measures were almost exclusively concerned with alleviating local effects. Considerable experience has been gained and many aspects of these problems are considered to be well under control now. Nevertheless, the behaviour in the environment of each particular pollutant can be complicated; the impact on human health and living conditions, on species of animal and plant life and on the material possessions of a community must all be considered. Of the hundreds of atmospheric pollutants, only a small number among the more significant can be mentioned here.

(a) Particulates. Under this heading, the main concern relating to airborne pollution is with particles whose sizes lie between 1,000 and 0.1 μm, emitted to atmosphere with effluent gases as grit, dust or smoke. Particles in the top part of this size range are normally referred to as grit. Such particles usually possess too much inertia to be drawn into the lungs and hence they pose little danger to health. Nevertheless, particles of large size fall through the atmosphere with significant speed and are readily detected if they hit a person's skin. Hence they can lead to vehement public complaint.

In the United Kingdom, particles in the size range from 75 to 1 μm are usually known as dust, while particles smaller than 1 μm are called fume or smoke. If such particles are present in the atmosphere, they are readily taken into the lungs when breathing and a significant proportion may be deposited there. They represent a very serious threat to health, especially so if the solid itself is insoluble, or toxic because of its chemical nature. While particles in the grit size range are relatively easy to remove from effluent gas streams, this becomes increasingly more difficult and expensive with smaller particles [29]. With extremely small particles, such as smoke, removal from a gas stream is scarcely practicable. Hence the only feasible approach is to avoid formation of such particles, by appropriate choice of fuel or conditions of combustion.

The effects on the environment of some particulate emissions have been widely reported, as have appropriate methods of control. These emissions include inorganic fluorides from brick and ceramic works, lead from lead smelters, dust from iron and steel works, in addition to ash, dust and soot from the combustion of many fuels.

(b) Gases. Emissions of sulphur dioxide and nitrogen oxides can no longer be regarded as simply local pollution issues since they appear to be primary causes of "acid rain" and associated problems affecting vast areas of the industrialised world. Indeed, many other gaseous substances, both organic and inorganic, appear to affect the natural balance between components of the atmosphere and therefore contribute to regional and global problems. Thus fewer pollutants can be thought of as having harmful effects simply within a short distance of their point of discharge.

As shown in Table 2, carbon monoxide is discharged to atmosphere in very large quantities, mainly in the exhaust gases of vehicles with petrol engines, but also from some chemical and metallurgical processes. In urban areas with high traffic densities, carbon monoxide concentrations can reach values in excess of World Health Organisation guidelines [1]. Vehicle drivers and traffic policemen may be temporarily affected [30]. However carbon monoxide emissions do not appear to have any long-term detrimental effects on the environment.

Many organic chemicals have a characteristic odour, even in very dilute concentrations. Emissions to atmosphere of such substances can cause considerable distress to people living in the locality of the source, although the amounts may otherwise pose no threat to health [31].

Effects of Atmospheric Pollution on Health

Among all the effects of pollutants, the one of most immediate interest and concern is the impact on human health. Each day an adult inhales a volume of air normally in the range 10 to 15 m^3. While the lungs provide a very effective means of absorbing oxygen into the blood stream, they provide similar facilities for any gaseous impurities in the air. Also, any small solid particles in the atmosphere can be drawn into the lungs with each breath and a proportion may be deposited on the lung surface. Occupational diseases such as silicosis, which can arise among workers subjected to high levels of dust, have been recognised for many years.

It is not easy to define an adverse effect on health due to air pollution. A list of possibilities may be drawn up, ranging from silent accumulation of chemicals or dust, through minor symptoms such as coughing, excess sputum, breathlessness, nasal irritation or eyes watering, then major signs like jaundice or changes in lung function, to serious illness, disability and death [32]. In studies of atmospheric pollution problems, the subtle effects and minor symptoms will be difficult to detect and may be a cause for debate as to whether they are actually occurring. The severe effects will be noted and agreed much more readily.

Methods to study the effects of airborne pollution on health fall into three main areas:
(i)   epidemiology,
(ii) animal toxicology, supported by short-term bioassays,
(iii) controlled clinical studies.

Epidemiology

Epidemiology of air pollution is the systematic study of naturally occurring associations between health effects and pollution [33]. Fruitful areas of research are data arising from the regular contact of workers in particular industries with specific airborne pollutants and also data from identifiable periods in which sections of the general population have been subjected to contaminated air.

The obvious advantage of such studies is that they are concerned with human beings and refer directly to atmospheric concentrations which have occurred in the past. A potential drawback is that it is not easy to prove that any apparent trends in ill-health have been caused by particular pollutants. In addition to the airborne contaminants, other variables which may affect the response of an individual are meteorological conditions, age, sex, tobacco smoking, levels of education and income, differences in activity, residence and nationality.

Table 1 was a striking example of an epidemiological study. The 4000 extra deaths which occurred in Greater London during a period of a few days in 1952 are a chilling reminder of the potentially lethal effects of atmospheric pollution. Most of these deaths occurred among the elderly, the late middle aged and the very young [30].

Animal Toxicology

Studies of animal toxicology have one marked advantage over epidemiological studies. This lies in the setting of carefully controlled conditions, hence of establishing the effects of different variables independently. Obviously the major disadvantage lies in the difficulty of extrapolating the data to man. Essentially two types of study are undertaken, one aiming to investigate what happens to a number of the animals at given doses of pollutant and the second aiming to determine the impact on bodily functions. Several different species of animals can be used which, compared with man, may have more or less than, or equal sensitivity to, particular toxins [34]. Differences in body metabolism can lead to wide variations in the effects of a pollutant. Certainly, any attempts to predict the effects on man from animal tests alone are open to considerable doubt. Nevertheless the results can be valuable in supporting and aiding the interpretation of data arising from epidemiological studies.

In order to support, or supplant, animal testing, especially in discovering the possible carcinogenicity of some pollutants, short-term bioassays have been developed [35]. These measure genotoxicity (DNA damage, mutations and chromosomal abnormalities) or neoplastic cell transformations. They are quicker and cheaper than animal tests but show good correlation with them.

Controlled Clinical Studies

Controlled clinical studies on human volunteers comprise the third method by means of which the effects of atmospheric pollutants on health can be investigated [36, 37]. In such studies the conditions can be determined in advance and controlled carefully. Clearly, however, severe and irreversible effects on the well-being of the volunteers must be avoided. Only the more subtle changes can be fully investigated, such as reaction time, behavioural functions and sensory responses. Then these have to be interpreted in conjunction with other data for predictions of long-term effects to be made.

Certainly clinical studies can be used to look at the impact of individual pollutants, or of chosen combinations. In contrast, the epidemiologist has to accept the possible synergistic (interactive) effects of all the pollutants which may have been present in the atmospheric conditions to which the human subjects have been exposed.

Air Quality Standards

Probably the best basic data on the effects of exposure to chemicals in the atmosphere have been derived from the experiences of workers on industrial plants. The data are in the form of Threshold Limit Values (TLVs), published, and reviewed annually, by the American Conference of Government Industrial Hygienists. These Threshold Limit Values refer to airborne concentrations of substances and represent conditions under which it is believed that nearly all workers may be repeatedly exposed, day after day, without adverse effect. Many of the values are based on the experiences of workers on industrial plants, thus constituting epidemiological data. The remainder are derived from human volunteer tests, animal tests or intelligent guesswork, starting with the values adopted for substances with similar chemical structure. In the United Kingdom, Threshold Limit Values have been used for several years as a basis for predicting the concentration of pollutants in the atmosphere to which the general public may be safely exposed. For many atmospheric contaminants, the Threshold Limit Value is divided by a safety factor commonly of the order of 30. The basis of this calculation is usually justified by attributing a factor of 3 to account for the difference between 8 hours during a working day spent on a plant and 24 hours daily exposure to general atmospheric pollution. An additional arbitrary factor of 10 is also used to give special protection to the very young, the old and the highly susceptible, in comparison with a fit and vigorous workforce for whom the Threshold Limit Values have been drawn up.

Many other countries have arrived at air quality standards by similar methods. In the United States, for example, national ambient air quality standards have been set for five major pollutants, namely suspended particulate matter, sulphur oxides, photochemical oxidants, hydrocarbons and nitrogen oxides. These form targets to enable individual states to develop suitable plans for controlling atmospheric emissions [38]. Primary standards are chosen as those necessary to protect public health, while secondary standards are aimed to protect public welfare, including effects on vegetation, visibility, crops, man-made materials, animals, economic values and personal comfort and well-being. The measures are designed to protect the health of the most sensitive group of individuals in the population, but not necessarily the most sensitive members of that group [39]. Much of the current research into the adverse health effects of airborne pollutants is concerned with particularly vulnerable people [36, 37, 40, 41].

Air Pollution as a Cause of Lung Cancer

Emissions to the atmosphere of chlorofluorohydrocarbons may cause a reduction in the ozone content of the stratosphere, leading to an increase in the rate of skin cancer. A higher rate of diagnosis of melanoma, a particularly virulent form of skin cancer, has been noted in recent years [42]. Other atmospheric contaminants may be contributory factors in the occurrence of cancers in other parts of the body. Nevertheless this section will be concerned only with cancer of the lung.

The possibility of an association between pollution of the atmosphere and the incidence of lung cancer in the population exposed to the pollutants was first realised in the nineteen-thirties. About 20 years later, considerable consternation was caused by a study which identified a 50% increased risk of lung cancer in the most polluted areas of Liverpool, when compared to rates in non-polluted areas [43]. This survey attributed about 50% of the deaths from lung cancer to cigarette smoking and approximately 37% to a factor almost completely absent in rural areas. Particular attention was drawn to the suspected carcinogen benzo(a) pyrene, a product of the incomplete combustion of fossil fuels, which was found to be present in Liverpool's atmosphere at concentrations approximately ten times higher than in the rural areas under investigation.

Benzo(a)pyrene is one of a range of polycyclic aromatic hydrocarbons which may be present, together with carbon and various inorganic substances, in the small particulate matter which constitutes smoke. There is no direct way of determining whether these polycyclic compounds are carcinogenic to man when inhaled, but certainly they have caused cancer when applied in massive doses to the skins of animals. The American Conference of Government Industrial Hygienists has the opinion that, regarding carcinogens, the only safe level of concentration in the atmosphere is zero.

It has often been said that the majority of human cancers are caused by environmental factors. Certainly the problem of lung cancer is one of wide concern and its relationship with airborne pollutants has been the subject of many epidemiological studies. However, even though it is generally agreed that the influence of air pollution is smaller than indicated by the conclusions of the 1955 study, there is still controversy about its importance as a contributory factor.

Research Methods

Researchers encounter a number of severe difficulties in this field, chiefly because many other factors relating to an increased risk of cancer are involved. Certainly cigarette smoking is the overwhelming cause of lung cancer [44] and any other contributory factors are prone to be masked by the effects of cigarettes. A simple comparison of lung cancer rates in urban areas with those in rural areas will be ineffective. It is now recognised that smoking is both more prevalent in urban locations [45] and also is subject to differences in smoking rates and the average age at which residents start to smoke [46].

Another significant influence in lung cancer rates is occupational exposure [47]. Concentrations of carcinogens in the vicinity of some industrial processing plants are likely to be much higher than in normal ambient air.

Therefore one would expect that workers near such plants would suffer higher rates of lung cancer than non-exposed workers. Since these plants are predominantly situated in urban rather than rural areas, this is a likely additional cause of the former area having a higher incidence of cancer. With both cigarette smoking and occupational carcinogens, there may be synergistic effects with carcinogens which are general airborne pollutants. Thus these factors may be interdependent and it may not be possible to establish their individual contributions to the overall lung cancer rate [48, 49, 50].

Table 3. Concentrations of benzo(a)pyrene in air at sites in central London, 1949–73, based on 24 hour samples aggregated for yearly periods

Period	Sampling site	Benzo(a)pyrene $\mu g/1{,}000\ m^3$
1949–51	County Hall	46
1953–56	St. Bartholomew's Hospital	17
1957–64	County Hall	14
1972–73	St. Bartholomew's Medical College	4

Reproduced from "Environmental Effects of Utilising More Coal" [52]

A further severe complication in the study of the causes of lung cancer arises from the long latent period [51]. From first exposure to a carcinogen, the development of a large tumour may take ten, twenty or thirty years. Hence the incidence of cancer should be related to the atmospheric pollution levels of previous eras, in which they may not have been measured accurately. In the intervening period considerable movement of the population, both with respect to place of residence and type of occupation, may have taken place. Table 3 [52] shows how the atmospheric concentrations of benzo(a)pyrene in Central London changed dramatically over a 25-year span. In this time, the London smog occurred and the ensuing 1956 Clean Air Act became fully operational.

Other factors which should be taken into account in an epidemiological study in this subject are that the probability of contracting cancer is dependent on age and that the level of pollution experienced in a particular zone of a large urban area may depend on the socioeconomic status of the residents in that zone [53].

Results

The broad aim of most studies in this field is to attempt to compare lung cancer rates in groups of people exposed to different levels of air pollution. Several investigations have been concerned with the effects of particular pollutants, released from a point source. Airborne emissions of arsenic, released for example from metal smelters, are said to be among the most likely to increase the lung cancer rate [54, 55]. An increase of approximately 15% in the lung cancer mortality figures was detected among the population subjected to the emissions, compared with those in similar areas remote from a smelter. Occupational exposure was unlikely to be the cause since women appeared to suffer an increased risk similar to the men. High rates of lung cancer were reported in two Scottish steel towns, including a cluster of 134 cases in the streets immediately surrounding one of the steel foundries [56]. Another series of studies showed a large number of cases of lung cancer in a community subject to air pollution by solvent vapours from a chemical plant [57]. The large number could not be explained by either cigarette smoking or lung cancer.

Some studies have investigated the health statistics of groups of people who have changed their country of residence, for example from England either to Australia, South Africa or New Zealand [44] and from various countries to the United States [46]. Such groups tend to have lung cancer rates which are in an immediate position between those prevailing in their country of origin and those in thir new country. The differences may be due to the changes in exposure to airborne pollutants which they have experienced. However it is difficult to make quantitative deductions of the alterations in the risk since their occupations, smoking habits and general lifestyle may be different from those normally characteristic of either their original or their new country.

One of the most thorough investigations into the link between lung cancer and general atmospheric pollution was based on a population of one million people in the United States [50]. The analysis was restricted to men who had lived in the same area for ten or more years and all results were adjusted for age and smoking habits. Two main groups of people were considered, those occupationally exposed to dust, fumes, gases or radiation and those not exposed because of their occupation. Disregarding the place of residence, it was found that occupationally exposed men suffered a 13.5% increased risk of lung cancer. Taking into account the type of residential area, the increase was 26% in a large metropolis, 18% in a smaller metropolitan area and 7% in a non-metropolitan area. Although these figures could be taken to indicate a synergistic effect between occupational exposure and the level of air pollution, it was said that a more likely explanation lay in the different types of occupational exposure in each area. When men with no occupational exposure were considered, much of the urban-rural difference disappeared and it was concluded that general air pollution had very little effect, if any, on the lung cancer rate. This finding has been supported by Doll and Peto [58]. Allowing for the falling concentrations of benzo(a)pyrene and associated combustion products in town air (as illustrated by Table 3) their contribution was thought unlikely to account for more than 1% of cases of cancer of the lung in the future.

Economics of Air Pollution

Earlier in this chapter, some incidents in the history of air pollution were described in which the sequence of events followed a common pattern. As emissions of a pollutant or set of pollutants increased so that a public outcry followed, suitable legislation was enacted forcing the emissions to be reduced or stopped. In general the aim of the legislation was to protect the well-being of the public by achieving a balance between the interests of the emitters and the receivers of the pollutants. In the UK, the Industrial Air Pollution Inspectorate uses judgement in interpreting "best practicable means" to control emissions to atmosphere, balancing the effects on the environment against the ability of industrial firms to bear the costs of control equipment [8]. However it is recognised that complete evaluations in monetary terms are seldom possible and experience has to be used in arriving at a decision.

Strategy and Costs of Air Pollution Control

The best approach for tackling a potential air pollution problem has been outlined previously [59]. The following possibilities should be considered in turn:
(1) prevent the process making the polluting substance, perhaps by a change in
     the choice of raw materials or by altering the process conditions;
(2) reduce emission of the polluting substance by its removal from the process gas
     stream;
(3) effect dispersal of the effluent gas into the atmosphere.
This order is correct both ideally and practically. Whilst the first step would solve
the problem at once, step 2 probably involves changing the form of the pollutant,
perhaps to a liquid phase solution or by chemical conversion to other substances.
Unless such streams can be usefully recycled to the process, the designer must be
able to dispose of them safely; this may simply lead to pollution of the environment in a different mode.

The third step in the above sequence should be considered last for a number
of reasons. Firstly effective dispersion may not be possible under all sets of meteorological conditions; atmospheric inversions may lead to a gradual increase in
pollutant concentrations in the vicinity of the source. Secondly there may be interactions between contaminants of the atmosphere, with unpredictable results.
Thirdly, as has become evident in recent years, widespread emissions of pollutants
may accumulate in the atmosphere sufficiently to cause damage over large regions
of the world, even when they appear to have been effectively dispersed from their
points of emission.

Because waste gas streams are low in density, they are often large in terms of
volumetric rate. Hence, in order to treat them for purposes of pollution control,
plant items of large cross-sectional area are needed. The addition of such items
to a process implies that extra power will be required to move the gases through
these items and hence a substantial increase in operating costs may result. Pollutants prior to emission to atmosphere are usually present in a gas stream only in
dilute concentrations and so it is not easy to devise effective treatments to remove
them. Mass transfer rates are often very small at low concentrations.

From these considerations it is clear that the costs involved in air pollution
control can be very large, both in terms of capital investment and also operating
costs. For a variety of industries, the equipment costs for gaseous effluent treatment often lie in the range from 5 to 15% of the entire plant capital costs [60].
The total expense of removing solid particulate matter from a gas stream can be
estimated reasonably accurately, if the particle size distribution and the required
removal efficiency are known [29]. With gaseous contaminants, the position is
more complicated, since the best method of treatment by step 2, and hence the
cost, depend on a variety of factors such as the solubility, chemical reactivity and
adsorptivity of these components. Avoidance or reduction of a problem using
step 1 (as, for example, stopping discharges of carbon dioxide from fossil-fuelled
electricity generating stations by changing to processes involving nuclear fission
or renewable energy sources) would often bring in complex issues of politics, as
well as costs. To solve most large-scale problems of atmospheric pollution, substantial increases in investment and operating costs are usually implied.

Costs of Damage

Assessing the total cost of the damage caused by the emission of an atmospheric pollutant is an extremely complex task. It is usually much more difficult than estimating the expense of reducing or stopping the emission. The harmful effects may include increased illness and earlier death of humans and other animals, contamination of surface waters with ensuing damage to fish, impaired growth of trees and other agricultural crops, corrosion of metals, deterioration of a variety of materials from stone to textiles and loss of amenity through malodour and reduced visibility. It is theoretically possible for a comprehensive research programme to provide approximate monetary values for many of these effects, although assigning a value to freedom from malodour and atmospheric haze would be especially difficult. Even allowing for the advances in scientific techniques of recent years, this is still an imposing task, since many current problems are being made even more complex by interactions between individual pollutants.

Another factor hampering progress in this field is the difficulty of relating the work done for one country to a different one. Variations in climate will affect the influence of a pollutant, as will the background levels of other pollutants. Effects on health, possessions and crops will depend on a wide variety of factors such as the racial characteristics of the people, their standard of living and occupations.

Probably the best study relating to the United Kingdom was published in 1972 [61]. (The figures should be increased approximately five-fold in order to give values in £s appropriate to the mid-nineteen eighties.) The total direct cost of air pollution in the UK was estimated to be about £410 × 10^6 per annum, including £5 × 10^6 for window cleaning, £42 × 10^6 to allow for corrosion of metals and preventive methods, £33 × 10^6 for damage to textiles, paper and similar materials, £195 × 10^6 for the effects on agricultural produce and £130 × 10^6 for direct damage to health. In addition, £780 × 10^6 was estimated for social or disamenity costs, including £510 × 10^6 for health, £100 × 10^6 for amenity and £170 × 10^6 for a number of other effects. Thus the total cost to the UK was assessed at the substantial figure of £1,190 × 10^6 per annum, equivalent to £21 per head of population. In 1985 values, the corresponding figures are approximately £6 × 10^9 per annum total, or £100 per annum per member of the population.

The uncertainties in such calculations are well demonstrated by estimates published by the Organisation for Economic Co-operation and Development [62]. After considering the economic value of the damage caused by sulphur dioxide emissions, including health effects, corrosion of materials, crop damage and acidification of waters, it was thought that the benefits of reduced emissions within the OECD European countries as a whole, would lie within the very wide range from £80 to £800 per tonne of sulphur dioxide. Health costs were assessed as the largest share of the total, but subject to the greatest uncertainty.

Perspective

For hundreds of years, pollution of the atmosphere has been a subject of considerable concern to people exposed to it. In the worst affected areas, it has caused substantial damage to property, as well as impairment of the quality of life through such effects as deposition of dirt, stabilisation of fogs and spreading of malodour. Even more seriously, it has damaged health and led to the early deaths of large numbers of people, especially in notorious episodes such as the London smogs of the nineteen-fifties.

However, in the last quarter of a century, the most pressing atmospheric pollution problems have changed in nature. Up to this period, virtually all of the damage caused by a pollutant occurred within a few kilometres of the emission source. The bulk of the atmosphere was regarded as a sink of almost infinite size so that pollutants which were dispersed into this could be regarded as having been safely disposed of. In recent years it has become clear that this view has been mistaken. Worldwide emissions of such compounds as sulphur and nitrogen oxides, hydrocarbons and chlorofluorocarbons, even if at dilute concentrations, are undergoing chemical changes within the atmosphere. They and the products of their reactions are being transported over large distances and are upsetting the balance of nature in many countries. (There is a parallel in the history of the treatment of liquid and solid wastes. For many years the oceans were regarded as if of infinite size for receiving toxic wastes. However, in 1972, many of the world's industrialised nations agreed to restrict the dumping in the deep seas of some of the more dangerous materials [63].)

In its legislation on pollution control, society has generally tried to strike a fair balance between the interests of the emitters of a pollutant and the recipients. However, even in the local pollution problems which predominated in past years, this has not been easy to achieve. Often the effects on health and on the environment may have been complex and difficult to assess. On the other hand, the treatment of large volumes of effluent gases may have been expensive, so that the imposition of very high emission standards may have threatened the viability of the underlying processes.

With regard to current problems concerning atmospheric pollutants which have an impact over large areas of the world, the difficulties of achieving a similar fair balance have been magnified enormously. In such problems, it is much more difficult to determine the links between individual pollutants and damage to the environment. Our knowledge of the long-range transport of gases and of their atmospheric reactions is far from complete. Nevertheless it is even more important than in the past that solutions are obtained within a reasonable time-scale. The worldwide cost implications of the "acid rain" problem appear to be increasing steadily each year and even when a fuller scientific understanding is reached, there are likely to be delays in reaching the necessary international consensus for appropriate action. Similarly, with the global issues of carbon dioxide emissions and depletion of stratospheric ozone and their possible effects on the world's climate, there is an urgent need for better understanding. Changes to the atmosphere which are occurring may take a long time to be slowed and reversed and the international co-operation required for corrective measures may be difficult to achieve.

References

1. Royal Commission on Environmental Pollution: 10th Report, Tackling Pollution – Experience and Prospects, HMSO, London (1984)
2. Holdgate, M. W.: A Perspective of Environmental Pollution, Cambridge University Press, Cambridge 1979
3. Summary of an Air Pollution Control Association Specialty Conference: J. Air Poll. Contr. Assoc. *34*, 799 (1984)
4. National Society for Clean Air: NSCA Reference Book, National Society for Clean Air, Brighton (1978), pp. 83–87
5. Hardie, D. W. F.: A History of the Chemical Industry in Widnes, Kynoch Press, Birmingham 1950, p. 19
6. Health and Safety Executive: Brief history of the Alkali Inspectorate 1864–1982, in: Industrial Air Pollution, Health and Safety 1982, HMSO, London 1984, pp. 18–20
7. Tomlinson, E. S.: "Best practicable means", in: Industrial Effluent Treatment, Vol. 2, Air and Noise (J. K. Walters and A. Wint eds.), Applied Science Publishers Ltd., London 1981, pp. 17–37
8. Health and Safety Executive: Industrial Air Pollution, Health and Safety 1983, HMSO, London 1985, pp. 2, 31
9. Royal Commission on Environmental Pollution: 4th Report, Pollution: Progress and Problems, HMSO, London 1974
10. Husselbee, W. L.: Automotive Emission Control, Reston Publishing Company, Inc., Reston, Virginia, USA 1984
11. Imai, M., et al.: J. Air Poll. Contr. Assoc. *35*, 103 (1985)
12. Buckly-Golder, D. H.: Acidity in the Environment, Report of Energy Technology Support Unit, HMSO, London 1984
13. House of Commons Environment Committee: Fourth Report from the Environment Committee, Acid Rain, Volume 1, HMSO, London 1984
14. The Watt Committee on Energy: Report Number 14, Acid Rain, The Watt Committee on Energy Ltd., London 1984
15. Royal Society Study Group: Long-term Toxic Effects; Final Report, The Royal Society, London 1978
16. Rotty, R. M.: CO_2 Assessment Program Contribution No. 81–23, Oak Ridge Associated Universities, Oak Ridge, Tennessee 1981
17. Emanuel, W. R., et al.: Modelling the circulation of carbon in the world's terrestrial ecosystems, in: Carbon Cycle Modelling, SCOPE 16 (B. Bolin ed.), Wiley, New York 1981
18. Clark, W. C., et al.: The carbon dioxide question: perspectives for 1982, in: Carbon Dioxide Review: 1982 (W. C. Clark ed.), Oxford University Press, Oxford 1982, pp. 3–44
19. Keeling, C. D., et al.: Measurements of the concentration of carbon dioxide at Mauna Loa Observatory, Hawaii, in: Carbon Dioxide Review: 1982 (W. C. Clark ed.), Oxford University Press, Oxford 1982, pp. 377–385
20. Mitchell, J. F. B.: Quarterly J. Roy. Meteor. Soc. *109*, 113 (1983)
21. Molina, M. J., Rowland, F. S.: Nature *249*, 810 (1974)
22. Fowler, D. et al.: Nature *297*, 383 (1982)
23. Franks, J.: Chemistry in Britain *19*, 504 (1983)
24. Kyte, W. S., et al.: Sulphur oxides control options in the UK electric power generation industry, in: Effluent Treatment in the Process Industries, The Institution of Chemical Engineers Symposium Series No. 77 (1983), pp. 39–48
25. Highton, N. G., Webb, M. G.: J. Ind. Econ. *30*, 49 (1981)
26. Ashenden, T. W., Mansfield, T. A.: Nature *273*, 142 (1978)
27. Howells, G. D., Kallend, A. S.: Chemistry in Britain *20*, 407 (1984)
28. House of Commons Environment Committee: Fourth Report from the Environment Committee, Acid Rain, Volume 2 (Minutes), HMSO, London 1984
29. Walters, J. K.: The removal of particulates from gases, in: Industrial Effluent Treatment, Vol. 2, Air and Noise (J. K. Walters and A. Wint eds.), Applied Science Publishers Ltd., London 1981, pp. 39–63
30. Logan, W. P. D.: Lancet *1*, 336 (1953)

31. Wint, A.: Odour control, in: Industrial Effluent Treatment, Vol. 2, Air and Noise (J. K. Walters and A. Wint eds.), Applied Science Publishers Ltd., London 1981, pp. 65–96
32. Higgins I. T. T.: J. Air Poll. Contr. Assoc. *33*, 661 (1983)
33. Goldsmith, J. R.: Arch. Environ. Health *18*, 516 (1969)
34. Miller, F. J., et al.: Environ. Health Persp. *52*, 169 (1983)
35. Albert, R. E.: Environ. Health Persp. *47*, 339 (1983)
36. Hackney, J. D., et al.: Environ. Sci. Technol. *18*, 115A (1984)
37. Hackney, J. D., Lynn, W. S.: Environ. Health Persp. *52*, 187 (1983)
38. Beachler, D. S., Joseph, G. T.: Environ. Progr. *3*, 44 (1984)
39. Padgett, J., Richmond, H.: J. Air Poll. Contr. Assoc. *33*, 13 (1983)
40. Colluci, A. V., Streiter, R. P.: Environ. Health Persp. *52*, 221 (1983)
41. Kleinman, M. T.: J. Air Poll. Contr. Assoc. *34*, 32 (1984)
42. Lee, J. A. H.: Epidemiol. Revs. *4*, 110 (1982)
43. Stocks, P., Campbell, J. M.: Brit. Med. J. *2*, 923 (1955)
44. Carnow, B. W.: Environ. Health Persp. *22*, 17 (1978)
45. Rantaneu, J.: Environ. Health Persp. *47*, 325 (1983)
46. Spiezer, E. S.: Environ. Health Persp. *22*, 22 (1978)
47. Maclure, K. M., Macmahon, B.: Epidemiol. Revs. *2*, 19 (1980)
48. Shephard, R. J.: Can. Med. Assoc. J. *118*, 379 (1978)
49. Doll, R.: Environ. Health Persp. *22*, 1 (1978)
50. Hammond, E. C., Garfinkel, L.: Prevent. Med. *9*, 206 (1980)
51. Friberg, L., Cederlof, R.: Environ. Health Persp. *22*, 45 (1978)
52. Lawther, P. J.: Carcinogens from coal and other sources, in: Environmental Effects of Utilising More Coal (Robinson, F. A. ed.), Royal Society of Chemistry, London (1980), pp. 14–18
53. McClellan, R. O.: Environ. Health Persp. *47*, 283 (1983)
54. Greaves, W. W., et al.: Am. J. Ind. Med. *2*, 15 (1981)
55. Goldsmith, J. R.: J. Environ. Pathol. Toxicol. *3*, 205 (1980)
56. Lloyd, O. L., et al.: Brit. J. Ind. Med. *42*, 475 (1985)
57. Capurro, P. U.: Clin. Toxicol. *14*, 285 (1978)
58. Doll, R., Peto, R.: J. Natl. Cancer Inst. *66*, 1191 (1981)
59. Walters, J. K., Wint, A.: The Chem. Engr. *315*, 751 (1976)
60. Ireland, F. E.: Gaseous and liquid effluent treatment versus recovery, paper 3A-2, in: Eurochem Conference, Birmingham 1980
61. Programmes and Analysis Unit of the Department of Trade and Industry and the UK Atomic Energy Authority: An Economic and Technical Appraisal of Air Pollution in the United Kingdom, HMSO, London 1972
62. Organisation for Economic Co-operation and Development: The Costs and Benefits of Sulphur Oxide Control: A Methodological Study, OECD, Paris 1981
63. Wint, A.: The disposal of toxic wastes, in: Industrial Effluent Treatment, Vol. 1, Water and Solid Wastes (J. K. Walters and A. Wint eds.), Applied Science Publishers Ltd., London 1981, pp. 263–291

Halogenated Hydrocarbons in the Atmosphere

P. Fabian

Max-Planck-Institut für Aeronomie
D-3411 Katlenburg-Lindau 3
Federal Republic of Germany

Summary

This report reviews the present knowledge about all halocarbon species that contribute to the stratospheric halogen budget. Since review articles published in earlier volumes of this handbook [13, 14] cover topics such as production processes, use patterns, the distribution in waters and biological systems, this chapter focuses on the atmospheric aspects, on budgets and the global distribution in the troposphere and stratosphere.

Introduction

Halogenated hydrocarbons or halocarbons play an important role in the photo-chemical processes of the atmosphere, since they constitute the source of halogen radicals which catalytically destroy ozone. A large variety of halocarbons is being released into the atmosphere by both natural and anthropogenic processes. Since natural sources are likely to remain essentially constant, we must realize that the anthropogenic contribution and hence the atmospheric halogen content are increasing with time.

Fully halogenated hydrocarbons such as trichloro-fluoromethane (CCl_3F or CFC-11) or dichloro-difluoromethane (CCl_2F_2 or CFC-12) are almost inert gases in the troposphere. Consequently, they accumulate and gradually diffuse into higher layers of the atmosphere, where they photolyse. Halogen atoms thereby released convert the "odd oxygen species", O_3 and O, to molecular oxygen O_2 through catalytic reaction cycles [1–4]. Molina and Rowland [5] first pointed out that the rapidly growing abundance of anthropogenically released CFC-11 and CFC-12 might thus cause a depletion of the Earth's ozone layer through chlorine atom catalyzed destruction in the stratosphere.

Partly halogenated hydrocarbons (hydrohalocarbons) such as methyl chloride (CH_3Cl), methyl chloroform (CH_3CCl_3), or chlorodifluoromethane ($CHClF_2$ or CFC-22), are less stable. They react with hydroxyl radicals and are thus largely removed by tropospheric OH-reactions. Only a fraction of the amounts emitted into the troposphere reaches the stratosphere and can thereby augment the source of halogen atom catalysts. The same holds for halocarbons with a double bond like perchloroethylene ($CCl_2=CCl_2$) or trichloro-ethylene ($CHCl=CCl_2$). The global release rates of CH_3Cl, CH_3CCl_3, and CFC-22 are, however, so large, that despite tropospheric removal mechanisms they nevertheless contribute considerably to the chlorine budget of the stratosphere.

Halocarbons that have been identified as atmospheric constituents are listed in table 1 in the order of their abundance. The 6 major halocarbons, methyl chloride, CFC-12, CFC-11, carbon tetrachloride (CCl_4), methyl chloroform, and CFC-22, listed in the upper part of the table, contain 88% of the organically bound chlorine present in the atmosphere. The other halogenated hydrocarbons, listed in the lower part of the table, hold the remaining 12%. However, the atmospheric lifetimes of CH_2Cl_2, $CHCl_3$, $CCl_2=CCl_2$ and $CHCl=CCl_2$ are so short, that they are almost entirely decomposed in the troposphere, their decomposition products being scavenged by precipitation. These substances will not be discussed in this review, which focuses on halocarbons that have an impact on the upper atmosphere.

The fully halogenated hydrocarbons, CFC-113, CFC-114, CFC-115, and CFC-13, together contain about 3% of the organically bound chlorine present in the atmosphere. Due to their extremely long life-times, however, the abundances of these manmade species will increase. Even for modest injection scenarios it can be foreseen that their contribution to the stratospheric chlorine budget will become more important in the future.

In recent years, the environmental concern about the accumulation of CFC-11 and CFC-12 in the atmosphere has prompted considerable research efforts,

Table 1. Global tropospheric abundance of halocarbons

Species	Trade No.	Average abundance (pptV)	Tropospheric burden (Mt)	Chlorine contribution (pptV)	Atmospheric lifetime (years)
CH_3Cl	40	617	5.2	617	2 – 3
CCl_2F_2	12	300	6.1	600	105 – 169
CCl_3F	11	175	4.0	525	55 – 93
CCl_4	10	145	3.7	580	60 – 100
CH_3CCl_3	140a	130	2.9	390	5.7– 10
$CHClF_2$	22	60	0.9	60	12 – 20
CF_4	14	67	1.0		10,000
CH_2Cl_2	30	32	0.5	64	0.5
$CHCl_3$	20	30	0.6	90	0.3– 0.6
$CCl_2=CCl_2$	1110	26	0.7	104	0.4
$C_2Cl_3F_3$	113	18	0.6	54	63 – 122
CH_3Br	40B1	15	0.2		1.7
$C_2Cl_2F_4$	114	11	0.3	22	126 – 310
$CHCl=CCl_2$	1120	8	0.2	24	0.02
C_2ClF_5	115	4.1	0.1	4	230 – 550
C_2F_6	116	4	0.1		500 –1,000
$CClF_3$	13	4	0.07	4	180 – 450
CH_3I	40I1	2	0.05		0.01
$CHCl_2F$	21	2	0.03	4	2 – 3
CF_3Br	13B1	1	0.02		62 – 112

[a] All figures refer to the 1980 level. The atmospheric burden was calculated assuming that the tropospheric consists fo 1.0×10^{44} molecules. References are given in the text.

and several review articles and agency reports about these two halocarbons have been published [6–12]. This chapter reviews the present knowledge about all halo-carbon species that contribute to the stratospheric halogen budget. Since review articles published in earlier volumes of this handbook [13, 14] cover topics such as production processes, use patterns, the distribution in waters and biological systems, this report will focus on the atmospheric aspects, on budgets and the global distribution in the troposphere and stratosphere.

The Major Halocarbons:
Sources, Sinks, Tropospheric Distribution and Trends

Methyl Chloride (CH_3Cl)

Methyl chloride is the most abundant halocarbon in the atmosphere and the only chlorocarbon with at least one proven natural source. Lovelock [15] and, independently, Grimsrud and Rasmussen [16] first reported the presence of CH_3Cl in the atmosphere as 1.14 and 0.53 ppbV (10^{-9} by volume), respectively. Despite this discrepancy both reports indicate higher concentrations of CH_3Cl for maritime air as compared to air of continental origin. Measurements in seawater [17] confirmed that the ocean, in particular the tropical ocean, could indeed be a sig-

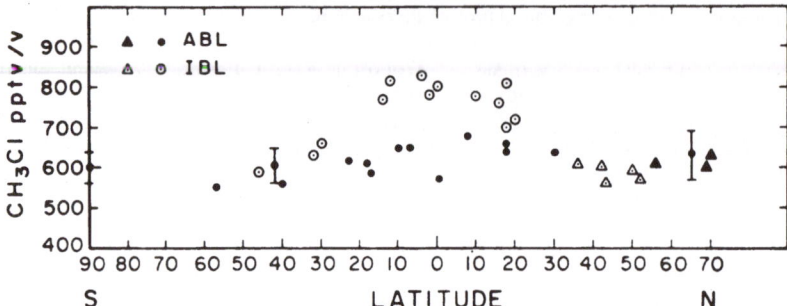

Fig. 1. Tropospheric distribution of CH_3Cl for 1978 Gametag flights made over the Pacific Ocean and Canada (Northwest Territory). Also included are CH_3Cl data obtained from ground sites at South Pole, Tasmania, and Alaska. Solid circles denote above boundary layer, Pacific Ocean; solid triangles denote above boundary layer, northern Canada; open circles denote in boundary layer, Pacific Ocean; open triangles denote in boundary layer, northern Canada; bars with solid circles denote ground sites; bars represent 1 standard deviation. After [18].

nificant source of the methyl chloride. Rasmussen et al. [18], during the 1978 GAMETAG flights made over the Pacific Ocean, found a distinct difference in the CH_3Cl measured in the atmospheric boundary layer versus the concentration above the boundary layer. The data suggest about 20% higher CH_3Cl mixing ratios in the boundary layer, and this phenomenon appears to be restricted to areas over the equatorial ocean (20° N to 30° S) with the maximum difference observed between 18° N and 14° S (Fig. 1). No such differences between the boundary layer and the free troposphere could be found over the continental United States, Canada, and Alaska.

From the concentrations of CH_3Cl measured in Pacific surface water Singh et al. [17] determined the oceanic source as 3.0×10^{12} g y^{-1}. Similar figures were obtained by Cicerone et al. [19], Yung et al. [20], and Graedel [21]. With the total atmospheric burden of CH_3Cl being 5.5×10^{12} g, an atmospheric turnover time of about 2 years results, which is in fairly good agreement with the estimated CH_3Cl residence time of about 2 to 3 years, due to OH attack (the OH concentration was assumed as 3 to 5×10^5 molecules per cubic centimeter) [17].

There is, however, a great deal of uncertainty with regard to the source strength and even the oceanic source relationship among the methyl halides, CH_3Cl, CH_3Br, and CH_3I. It is not known, for instance, whether CH_3Cl is formed by direct biosynthesis, or whether it is simply a product of a substitution reaction of Cl^- on CH_3I in seawater, as suggested by Zafiriou [22]. Based on this process, $CH_3I + Cl^- \rightarrow CH_3Cl + I^-$, Watson et al. [23] calculated the marine source of CH_3Cl ranging between 1 and 8×10^{12} g y^{-1}.

The natural oceanic production contributes about 85% to 90% to the total global source of methyl chloride, while 10% to 15% are likely to be due to other, mainly anthropogenic processes [24]. Rasmussen [25] suggested, based upon elevated CH_3Cl mixing ratios observed over areas experiencing tropical slash and burn agriculture such as Kenya, that biomass burning may be a significant source. The global source strength of this process was estimated by Crutzen et al. [26] ranging between 0.2 to 0.4×10^{12} g y^{-1}, about one order of magnitude lower than the oceanic emission rate.

Direct industrial emission of methyl chloride is thought to be small. For example, total production in 1973 was 0.35×10^{12} g, but little of this was released into the atmosphere [23]. However, measurements made by Singh et al. [17] suggest that a significant urban source does exist. Maximum CH_3Cl mixing ratios of 2,200 and 3,800 pptV (10^{-12} by volume) were observed in Lisbon and in the Los Angeles basin, respectively, about 3 to 6 times the natural background level. Automobile exhaust, the burning of PVC waste, and heterogeneous reactions occurring on the surface of smog particles, have been suggested as likely sources. The total of these urban sources may range between 0.15 and 0.6×10^{12} g y^{-1} [23]. CH_3Cl levels in excess of the background mixing ratio were also measured in the vicinity of volcanic eruptions [18, 27, 28], but for the global budget the volcanic source is likely to be negligible. Another possible natural source, i.e., the production by microorganisms through direct metabolism [29] cannot be assessed yet until more insight is gained into this process.

The following table summarizes the sources of methyl chloride.

Source	Strength (Mt per year)
Marine	1 –8
Biomass burning	0.2 –0.4
Urban anthropogenic	0.15–0.6
Microorganisms	?
Total	1.35–9.0

The main sink of CH_3Cl is due to the attack by hydroxyl radicals in the troposphere. The global removal rate through this process was estimated by Logan et al. [30] as 5.2×10^{12} g y^{-1} yielding a lifetime of methyl chloride of 1 year. This was, however, based on tropospheric OH concentrations ranging between about 2×10^5 and 2×10^6 cm^{-3}. Model calculations related to the global increase of the methyl chloroform abundance suggest that the OH concentrations have to be reduced by a factor of about 2 (see fig. 6), which would then halve the calculated global sink of CH_3Cl yielding a lifetime of 2 years.

The global distribution of CH_3Cl in the troposphere appears to be fairly uniform. Latitudinal surveys carried out by Singh et al. [17], between 60° N and 42° S, and Rasmussen et al. [18], between 70° N and the South Pole, showed no significant variations of the CH_3Cl mixing ratios with latitude. Sing et al. found an average of (611 ± 84) pptV and (615 ± 103) pptV for the northern and southern hemispheres, respectively; the corresponding figures given by Rasmussen et al. are (626 ± 40) pptV for both hemispheres, with slightly higher levels in the tropics.

Khalil and Rasmussen [31] report indications for a seasonal variation of methyl chloride. At 45° N, springtime mixing ratios were found to be $(7.5 \pm 3.5)\%$ higher than those observed during other seasons. There is, however, no indication for a longterm trend of the CH_3Cl abundance [18].

CFCl₃(CFC-11) and CF₂Cl₂ (CFC-12)

There is no doubt that the present atmospheric abundance of CFC-11 and CFC-12 is due to man's activities. Only anthropogenic sources are known with the possible exception of trace amounts released in volcanic eruptions [32]. These substances were invented in the 1930's, and CFC-12 was introduced at that time as a refrigerant. After World War II the aerosol industry developed, and CFC-11 achieved a widespread usage. Together with CFC-12 it became a major aerosol propellant and foam blowing agent. Besides their convenient thermodynamic properties, the main advantages for the use of these halocarbons are their chemical stability, non-flammability, and lack of toxic effects. The present emission rate is of the order of 250,000 and 330,000 t y^{-1} for CFC-11 and CFC-12, respectively. The total burden accumulated by 1980 was 4.3 Mt of CFCl₃ and 6.2 Mt of CF₂Cl₂ [33]. Until 1974, the global emission of both CFC-11 and CFC-12 into the atmosphere increased by more than 10% every year. It has slightly decreased since then levelling off at the rates quoted above (see Fig. 2). The American Chemical Manufacturers Association (CMA) have recently revised the figures of the global production and release of CFCl₃ and CF₂Cl₂, since production rates in the Eastern Block countries were obviously larger than previously assumed.

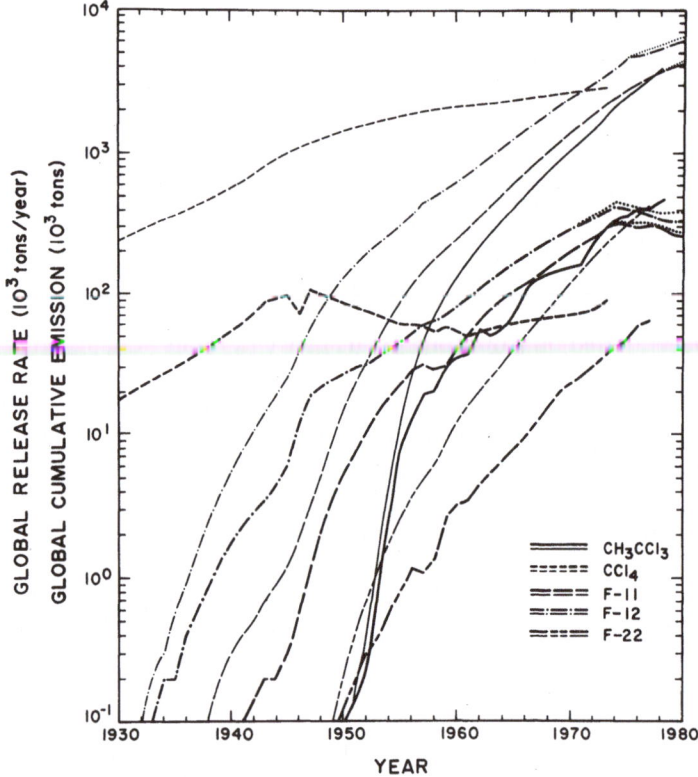

Fig. 2. Global release rates (heavy) and global cumulative emissions (thin) of the major anthropogenic halocarbons. Sources: CFC-11 and CFC-12 [33, 34], CFC-22 and CH₃CCl₃ [35], CCl₄ [36]. Recently updated emission rates of CFC-11 and CFC-12 [34] are shown by dotted lines.

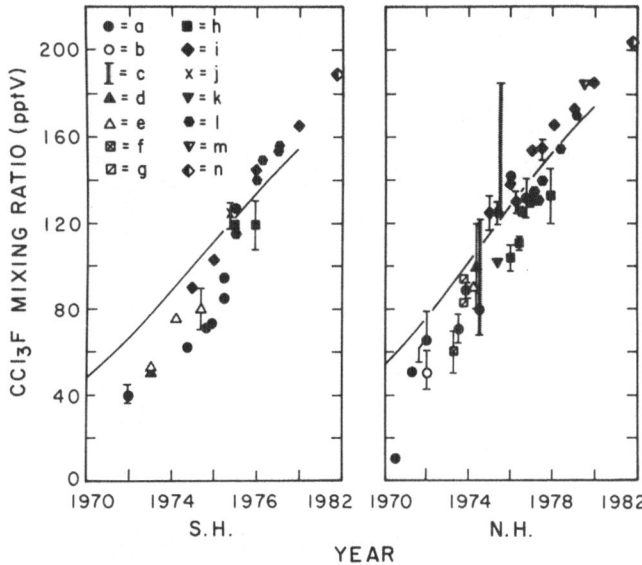

Fig. 3. Measurements of CFCl₃ in the northern and southern hemispheres as a function of time. Each symbol represents data from a different group of investigators: a (30°–90°), b (0–30°), c [41–44, 37, 10]; d (30°–90°), e (0–30°) [45–48]; f [49, 50]; g [51]; h [52, 53, 17]; i [16, 54–56, 38]; j [57]; k [58]; l [59]; m [60]; n [40, 61]. The solid line shows the result of model calculations. This figure is an update of a survey prepared by Logan et al. [30].

The revised emission rates for CFC-11 and CFC-12 are 270,000 and 390,000 t y^{-1} for 1980 [34] (see Fig. 2).

No efficient tropospheric removal processes have been identified, and it is very unlikely that a significant tropospheric sink will be found in the future [12]. The major removal process for both substances is photolysis in the stratosphere, which determines the atmospheric lifetimes of CFC-11 and CFC-12. These have been determined by a number of research groups using different sets of transport parameters, and range between 55 and 93 years for CFC-11 and between 105 and 169 years for CFC-12 [12].

The global distributions of CFCl₃ and CF₂Cl₂ have been thoroughly reviewed (see for instance [10, 12, 13, 30]). Since about 90% of the release takes place in the northern hemisphere, a hemispheric asymmetry exists. However, while during the time of rapid growth of the global release, at the end of 1971, CFCl₃ background mixing ratios observed in the northern hemisphere were almost twice those found over the southern hemisphere [37], the N/S ratio has decreased to about 1.1 in 1980 for both CFC-11 and CFC-12 [12, 38]. In both hemispheres CFC-11 and CFC-12 mixing ratios are increasing with time (Figs. 3 and 4). According to Khalil and Rasmussen [39] the annual growth rates of the CFCl₃ abundance have been decreasing from more than 15% in 1975 to 5%/year in 1980, those for CF₂Cl₂ from 10%/year in 1976 to 5%/year in 1980. Average mixing ratios measured during May, 1982 at 70° N are (205.7±0.4) and (353.8±0.7) pptV with interhemispheric ratios being (1.08±0.01) and (1.07±0.01) for CFC-11 and CFC-12, respectively [40].

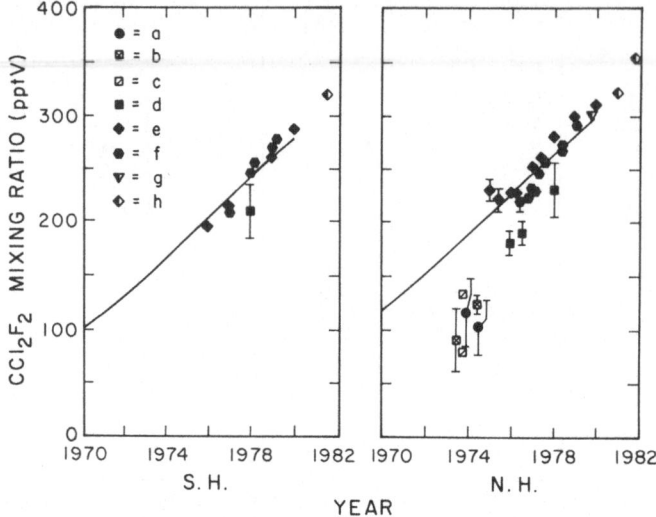

Fig. 4. Measurements of CF_2Cl_2 in the northern and southern hemispheres as a function of time. Each symbol represents data from a different group of investigators: a [41–44, 37, 10]; b [49, 50]; c [51]; d [52, 53, 17]; e [16 54-56, 38]; f [59], g [60]; h [40, 61]. The solid line shows the result of model calculations. This figure is an update of a survey prepared by Logan et al. [30].

Model calculations relating the stratospheric decomposition and the published release rates with the global increase of CFC-11 and CFC-12, reveal some discrepancies, which do not allow, however, for tropospheric loss processes [30]. The discrepancies between model results and recent background measurements are less than 10% [12]. They become smaller when revised CFM emission data are used.

In cities and industrial areas CFC-11 and CFC-12 mixing ratios can often significantly exceed the background levels [62, 63]. Volume mixing ratios in the ppb-range (10^{-9}) were observed in cities like Hamburg, Oslo, and Vienna [64]. Accordingly these substances can be used as tracers of urban air masses [41, 65].

Carbon Tetrachloride (CCl₄)

The atmospheric budget of carbon tetrachloride has been a subject of considerable uncertainty. Based on their first measurements of this constituent, Lovelock et al. [37] and Lovelock [43] suggested that CCl_4 must have a natural source. Their suggestion was based on two observations: (i) the global emission of CCl_4 seemed insufficient to account for the (71.2 ± 6.8) pptV measured in 1971/72 and (ii) the global distribution of CCl_4 appeared more uniform than that of CFC-11. By reassessing worldwide production-emission relationships, Singh et al. [66] could show, however, that the atmospheric abundance and distribution of CCl_4 are consistent with cumulative anthropogenic emissions. Graedel and Allara [67] argue that sources other than direct anthropogenic release are very unlikely.

The pattern of industrial CCl_4 usage has undergone significant changes over the last four decades. Before 1950, the major world market was the United States, where it was primarily used as an industrial solvent, dry cleaning agent, fire extinguisher and grain fumigant. In western Europe and Japan, however, no significant amounts of CCl_4 were used during that time. Since 1950, the world production of CCl_4 has kept pace with the production of chlorofluoromethanes, for which it is the principal starting material. Although the US production increased nearly tenfold between 1940 and 1972, the fraction emitted into the atmosphere declined from 84% to 7% of the total production [66]. Consequently, the worldwide emission showed a slight decline between 1950 and 1960, with a similarly slight increase thereafter (Fig. 2), which is obviously due to non-US sources [36, 66].

Large uncertainties still exist about the global emission of CCl_4 into the atmosphere, since production figures are unavailable for many countries including the USSR and China, and even existing production data do not allow for an accurate assessment of the actual emission. With an average abundance of 148 and 142 pptV, measured during 1980 on the northern and southern hemispheres, respectively, the global tropospheric burden of CCl_4 was 3.7 Mt at that time. The extrapolation of the cumulative emission curve in Fig. 2 yields exactly the same value. Thus, in view of the atmospheric lifetime of CCl_4, estimated to range between 60 and 100 years, it appears that the emission data were underestimated by about 10%. According to a recent study by Simmonds et al. [68], the lower and upper limits of the cumulative world emission of CCl_4, at the end of 1980, were 3.427 Mt and 4.036 Mt, respectively.

The emission curves also reveal that 65% of the total CCl_4 released into the atmosphere was emitted before 1960, whereas 85% of the CFC-11 was released after 1965. This additional time available for global dispersion of CCl_4, and its slow present release explain why, at the end of 1971, Lovelock found CCl_4 almost homogeneously distributed, while CFC-11 background mixing ratios on the northern hemisphere were nearly twice as high as on the southern hemisphere [66].

The sink mechanisms of CCl_4 have been assessed by Galbally [36]. According to this study, the main mechanism for its loss from the atmosphere is UV-photolysis in the stratosphere, which accounts for an atmospheric lifetime of 59 years. Tropospheric processes, such as gas-phase reactions, hydrolysis in water, and adsorption and decomposition in biological systems, are negligible; they account for lifetimes of 1,000, 143, and 500 years, respectively. Singh et al. [66] quoted a realistic estimate of the CCl_4 lifetime in the range of 60 to 100 years. With an average of 75 years they calculated the cumulative atmospheric build up of CCl_4 to be about 1.9×10^6 tons in 1973, which is about 75% of the worldwide cumulative emission at that time, and sufficient to account for the measured atmospheric abundance of carbon tetrachloride during 1973.

Worldwide measurements of CCl_4 are displayed in Fig. 5. Although inconsistencies in the results of different research groups exist, an increase of the CCl_4 mixing ratios is clearly noticeable for the 10 years period covered by this survey.

The N/S ratio for CCl_4 is close to unity. From aircraft measurements in the upper troposphere, between 80° N and 60° S, Tyson et al. [69] found an average of 1.01 for 1976. At the end of 1975, measurements aboard a research vessel, carried out between 60° N and 40° S, yielded 1.03 [17]. On the continent, where local

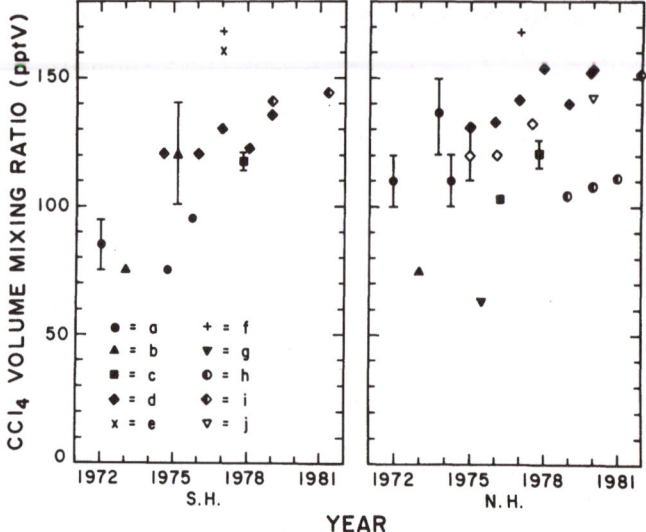

Fig. 5. Measurements of CCl$_4$ in the northern and southern hemispheres as a function of time. Each symbol represents data from a different group of investigators: a [41–44, 37, 10]; b [45–48]; c [52, 53, 17]; d [16, 54–56, 38]; e [57]; f [69]; g [58]; h [63]; i [40, 61]. This figure is an update of the WMO-NASA survey [121].

or regional pollution is likely to interfere, Rasmussen et al. [38] found values varying betwen 1.04 and 1.25, for the periode covering the years 1975 to 1979. Since CCl$_4$ often shows elevated concentrations in polluted areas [36, 62, 63], measurements made even far away from the source regions might occasionally yield mixing ratios higher than the background level, due to the long lifetime of CCl$_4$.

Methyl Chloroform (CH$_3$CCl$_3$)

Methyl chloroform, manufactured primarily for use as an industrial degreasing and dry cleaning solvent, has been released into the atmosphere in steadily increasing amounts since the early 1950's [35, 70]. Annual injection rates and cumulative emissions are shown in Fig. 2. Apparently the present annual release rate is more than twice that of CFC-11 or CFC-12, and its average annual growth after 1970 has been 16%/year. There are no natural sources of CH$_3$CCl$_3$ [67].

Like methyl chloride methyl chloroform is decomposed through the attack of hydroxyl radicals in the troposphere. Earlier estimates of the corresponding atmospheric lifetime for CH$_3$CCl$_3$ were affected by erroneous laboratory rate constants and uncertainties with respect to the global OH distribution. Published results, as reviewed by Makide and Rowland [71], vary between 1 year and 11 years. From the analysis of the global atmospheric distribution and the trend, a lifetime of (6.9 ± 1.2) years [71] or $(7 \pm^3_1)$ years [72] appears likely. There are, however, still large uncertainties because release rates provided by Neely and Plonka [70] do not account for the release in Eastern Block countries.

Since decomposition by OH attack is the only significant sink for CH$_3$CCl$_3$, it was suggested that the global distribution and its changes with time may provide information on the global OH distribution [30]. Measurements of methyl

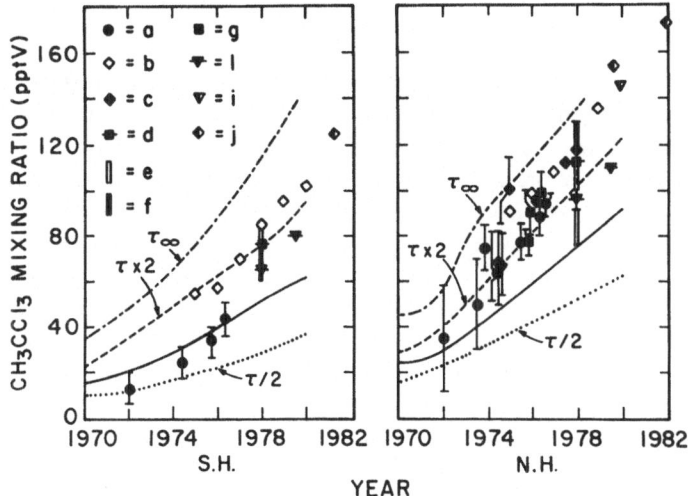

Fig. 6. Measurements of CH_3CCl_3 in the northern and southern hemispheres as a function of time. Each symbol represents data from a different group of investigators: a [43, 74]; b [38]; c [16, 55, 56]; d (hemispheric mean); e (range, 0–30°); f (range, 30°–60°) [17]; g [52, 53]; h [75]; i [60]; j [40, 61]. The lines show model results as discussed in the text. This figure is an update of a survey prepared by Logan et al. [30].

chloroform in the northern and southern hemispheres as a function of time are presented in Fig. 6 together with model results, based upon various assumptions concerning the OH distribution. The data show large discrepancies with are likely to be due to absolute calibration problems (see for instance [12, 71]). Furthermore, some of these data may not reflect background mixing ratios but rather show the influence of pollution sources. Nevertheless, there is no doubt that CH_3CCl_3 mixing ratios on both hemispheres have been growing by a factor of 4 from 1970 to 1980. Measurements made during November 1981 between 30° and 40° S [61] and in May 1982 at 70° N [40], with 124 pptV and 175 pptV, respectively, show that the growth of the CH_3CCl_3 abundance continues.

The solid model curves in Fig. 6 [30] correspond to an OH distribution with number densities varying between about 2×10^5 and 2×10^6 cm^{-3}. The dashed curves result from multiplying these by 0.5, while the dotted curves are for OH-concentrations multiplied by 2. The dot-dash curves show results for the absence of OH, i.e. an infinite tropospheric lifetime of CH_3CCl_3. Neither of these model curves gives a perfect fit to the observed data. It appears though that with reduced OH concentrations the agreement is best. The same holds for the interhemispheric ratios, which have been decreasing from about 1.7 in 1976 to 1.4 in 1980 [30].

The modelling work by Logan et al. [30] indicates that methyl chloroform is removed from the atmosphere mainly in the tropical troposphere, which accounts for about 70% of the global sink. It is interesting to note that the transition from the northern hemisphere mixing ratio to the southern hemisphere level takes place almost entirely within the tropics, whereas the background mixing ratios in the extratropical regions are nearly constant with latitude, as was shown by latitudinal surveys made by Singh et al. [17], Makide and Rowland [71] and Rasmussen and Khalil [73].

CHF$_2$Cl (CFC-22)

CFC-22, another man-made halocarbon, is primarily used in refrigeration and as a foam blowing agent. Its global production and release rates have been assessed by Jesson [35]. The data show that the injection of CFC-22 into the atmosphere will have reached the present release rate of either CFC-11 and CFC-12 by about 1988, if the present average increase of about 16%/year will persist (Fig. 2).

Since CFC-22 contains a hydrogen atom, it reacts with OH radicals. Like for methyl chloroform, this sink is sensitive to the global OH distribution. The atmospheric life-time of CFC-22 was estimated to vary between 12 years [12] and 20 years [76]. Accordingly, since atmospheric CFC-22 was first measured in 1979 by Rasmussen et al. [77], a pronounced increase of its abundance was found in both hemispheres. By analyzing ambient air samples stored in calibration tanks with inert internal surfaces, atmospheric mixing ratios could be traced back to April 1976 [78]. These data, displayed in Fig. 7 together with more recent results, show a consistent increase of 11.7%/year. Khalil and Rasmussen [78], however, point out that measured mixing ratios in the atmosphere correspond to an excess of about 26% over the cumulative anthropogenic emission as displayed in Fig. 2. They conclude that the past industrial release rates have been underestimated.

Indeed, with an average abundance of 65 and 55 pptV, measured during 1980 on the northern and southern hemispheres, respectively, the global tropospheric burden of CFC-22 was 0.9 Mt at that time. A rough extrapolation of the cumulative emission curve in figure 2 yields exactly the same value. In view of the atmospheric life time given above it would thus appear that the emission data are underestimated by at least 20%. Latitudinal surveys carried out during 1978 and 1979 indicate that the transition from the northern hemisphere mixing ratios to the southern hemisphere levels take place, like for methyl chloroform, almost entirely within the tropics [77, 73].

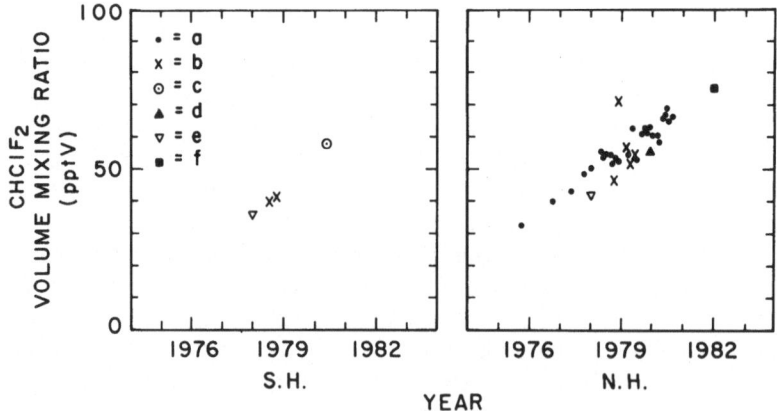

Fig. 7. Measurements of CHF$_2$Cl in the northern and southern hemispheres as a function of time. Each symbol represents data from a different group of investigators: a [78]; b [77]; c [61]; d [60]; e [73]; f [40].

The Stratospheric Distribution of the Major Halocarbons

In the stratosphere halogenated hydrocarbons are decomposed by UV photolysis and by reactions with metastable O('D) atoms. For molecules containing hydrogen the OH attack is important too (Absorption cross sections and reaction rate coefficients are compiled in ref. [79]). Hence, halocarbon mixing ratios decrease with increasing altitude, this decrease depending on the decomposition rate and the upward flux of the species. Under steady-state conditions, the photochemical loss is compensated by transport from below.

By far most of the data on the stratospheric distribution of halocarbons have resulted from the analysis of air samples collected aboard aircraft and balloon platforms. The analytical techniques applied, i.e., gas chromatography and mass spectrometry, are essentially the same as those applied for tropospheric studies. Both grab sampling [59, 80], also in connection with a compressor [40, 58, 60, 61, 69, 81], and cryogenic sampling techniques [51, 73, 82–84] have been applied successfully. Only very few data have so far been obtained by means of other methods such as infrared spectroscopy [85, 86].

In the following survey data obtained by different investigators are displayed such as to get mixing ratio/height profiles. Each symbol represents data from a different group, the time of the measurement is also given in the legends of the figures. Unless otherwise indicated, the data represent northern midlatitude conditions. No attempt was made to reproduce error bars, since error estimats published by the different groups of investigators are difficult to evaluate such as to provide a basis for a meaningful comparison, and in many cases error estimates are not given at all. The relative errors for the individual profiles are probably of the order of $\pm$ (10–20%). The absolute calibration may, however, add systematic errors which cannot be assessed from the literature.

Methyl Chloride (CH₃Cl)

Figure 8 shows the vertical distribution of methyl chloride. Surface measurements (a) aircraft, (b, f–h) and balloon data, (c–e) yield a consistent profile, with mixing ratios decreasing from 617 pptV found in the troposphere, to 5–10 pptV at 34 km. The tropospheric data from the southern hemisphere (g) do not significantly differ from their northern counterparts. Above 20 km, only 3 balloon profiles exist so far, which are, despite considerable scatter of the individual data points, in reasonable agreement with model results [88, 89].

CFCl₃ (CFC-11)

The vertical distribution of CFC-11 is displayed in Fig. 9. Again, ground-based, aircraft, and balloon data shape a consistent profile, but two types of scatter are noticeable: (i) in the stratosphere some points, in particular those of profile a, show considerably higher mixing ratios than the rest of the data. This is most likely due to the fact that these measurements were made at 32° N, where interferences with the tropical circulation often occur. Because of intensive upwelling motions in the tropics which counteract the photochemical losses, mixing ratios of source gases such as CFC-11 and CFC-12 decrease more gradually there than

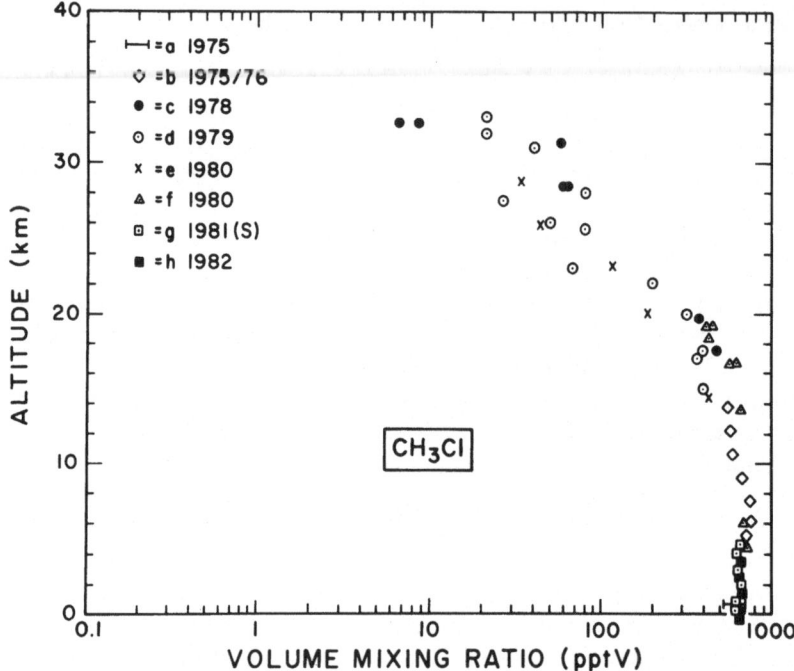

Fig. 8. The vertical distribution of CH_3Cl in the atmosphere. Each symbol represents data from a different group of investigators. The time of measurement is also given in the figure. Unless otherwise marked by (S) behind the year, all data originate from the northern hemisphere: a [17]; b [87]; c [88]; d [89]; e [90]; f [60]; g [61]; h [40]

at middle and high latitude [59]. (ii) A general increase of the CFC-11 abundance is noticeable at almost every height level. In the troposphere, the buildup from 100 pptV in 1974 [43] to 205 pptV in 1982 [40] is quite obvious. Despite the scatter and some discrepancies, some of which are probably due to variations of the tropopause height, the buildup is noticeable up to almost 30 km altitude. In the upper portions of the profiles, the data show a larger scatter due to the low mixing ratios, which are close to the detection limit.

Models generally predict a slower falloff of the CFC-11 abundance with height than that actually measured in the stratosphere [12]. Froidevaux and Yung [94] showed that a substantial part of this discrepancy is removed, when O_2 absorption cross sections in the important 200 to 220 nm spectral window are reduced, this reduction being suggested on the basis of recent measurements of the solar UV fluxes between 32 and 39 km altitude [95, 96]. A further improvement is achieved by including second-order effects of stratospheric vertical motions in the transport schemes of one- and two-dimensional models [97].

CF_2Cl_2 (CFC-12)

The same features discussed for CFC-11 can also be found in the vertical profiles of CFC-12, displayed in Fig. 10. Compared to CFC-11, CFC-12 contains one chlorine atom less, and is thus more stable with respect to UV photolysis [98, 99].

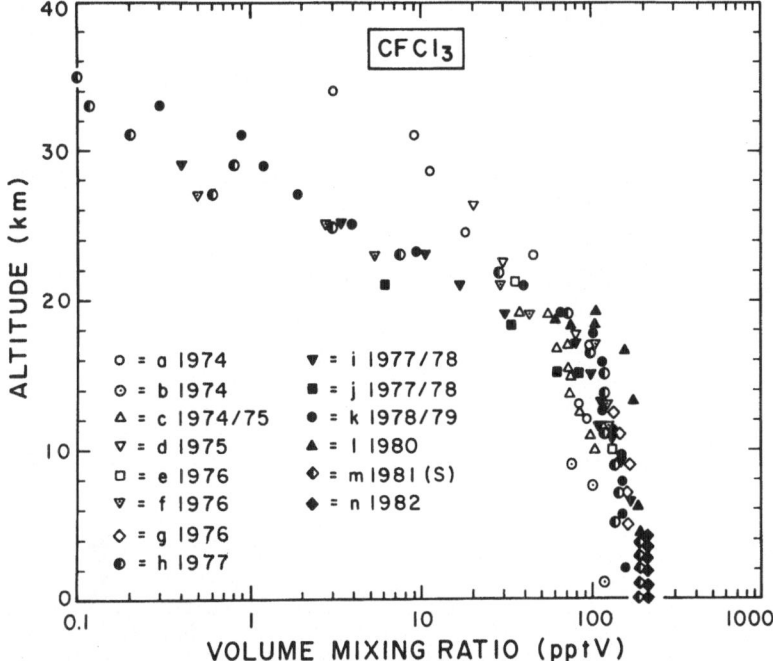

Fig. 9. The vertical distribution of CFCl₃ in the atmosphere. Each symbol represents data from a different group of investigators. The time of the measurement is also given in the figure. Unless otherwise marked by (S) behind the year, all data originate from the northern hemisphere. a [51]; b [43]; c [58, 91]; d [80]; e [82]; f [59]; g [81]; h [83]; i [59]; j [92]; k [93]; l [60]; m [61], n [40]

Its mixing ratio decreases less rapidly in the stratosphere, and chlorine atoms are thus released at higher altitudes than from CFC-11. As with CFC-11, model computations yield higher mixing ratios than observed in the stratosphere, this discrepancy becoming smaller when reduced O_2 absorption cross sections in the Herzberg continuum, between 200 and 220 nm, are used [94].

Carbon Tetrachloride (CCl₄)

Until recently, no measurements of CCl_4 above 21 km were available. There are quite a number of ground-based and aircraft measurements, the results of which are displayed in Fig. 11. Some of the scatter they show is likely to be due to experimental errors, but a tendency for a general increase withing the 9 years covered by the data is clearly noticeable. The two tropospheric profiles measured by Rasmussen et al., during November 1981 on the southern (p) and during May 1982 on the northern hemispheres (q), show average mixing ratios of (145 ± 1) and (150 ± 2) pptV, respectively. This 3% increase within 6 months reflects both the global buildup and the interhemispheric difference.

Aircraft samples taken between 19 and 21 km show a latitudinal dependence similar to that observed for CFC-11 and CFC-12: the stratospheric decrease of the mixing ratio is slower in the tropical region (h, m) than at midlatitudes (g, l), and it appears to be even faster at high latitudes (f, k). Based on a measured ver-

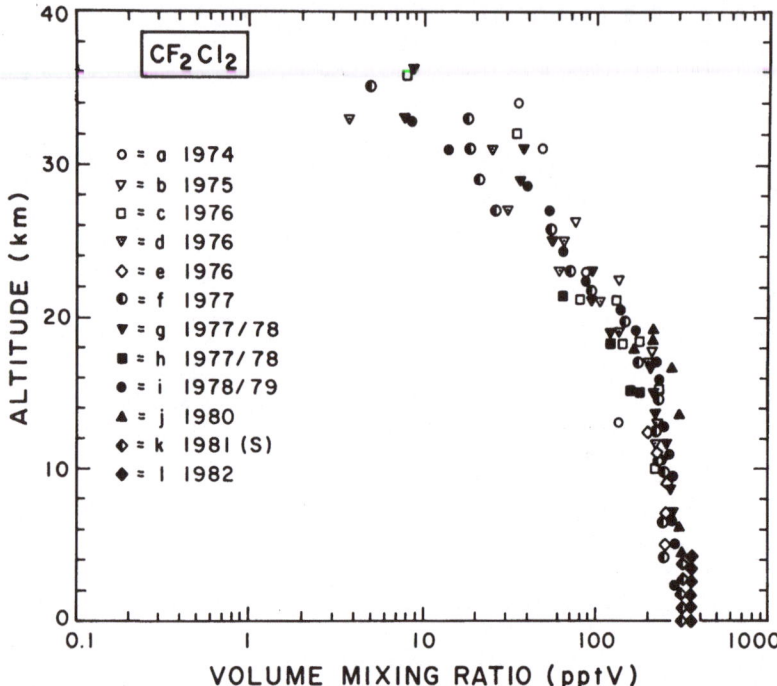

Fig. 10. The vertical distribution of CF_2Cl_2 in the atmosphere. Each symbol represents data from a different group of investigators. The time of the measurement is also given in the figure. Unless otherwise marked by (S) behind the year, all data originate from the northern hemisphere. a [51]; b [80]; c [82]; d [59]; e [81]; f [83]; g [59]; h [92]; i [93]; j [60]; k [61]; l [40]

tical profile up to about 20 km (d), Krey et al. [58] made an extrapolation for the height region above (dashed line).

As model calculations show [106], this extrapolated part of the profile is certainly wrong, since CCl_4 falls off much more rapidly above 20 km. Recently, from cryogenically collected balloon samples, a first complete stratospheric profile of CCl_4 could be measured (r). It confirms the rapid falloff in the stratosphere, with mixing ratios decreasing to about 0.1 pptV at 26 km.

Methyl Chloroform (CH_3CCl_3)

Figure 12 shows the vertical distribution of methyl chloroform, as given by the existing data. From cryogenically collected balloon samples Fabian et al. [90] had obtained a tentative stratospheric profile showing a decrease of the CH_3CCl_3 mixing ratio to about 1 pptV at 23.3 km (d). This measurement suffered from considerable decay problems in the sampling tubes. Recently, with a newly designed system, a complete profile could be obtained, with mixing ratio decreasing from about 100 pptV at 10 km to 0.5 pptV at 26 km altitude (g). As was also found for other halocarbons, the modelled profile [106] falls off less rapidly than the measured one. The fast growth rate of the CH_3CCl_3 abundance is noticeable from the comparison of the two tropospheric profiles e and f measured 6 months apart on the southern and northern hemispheres, respectively.

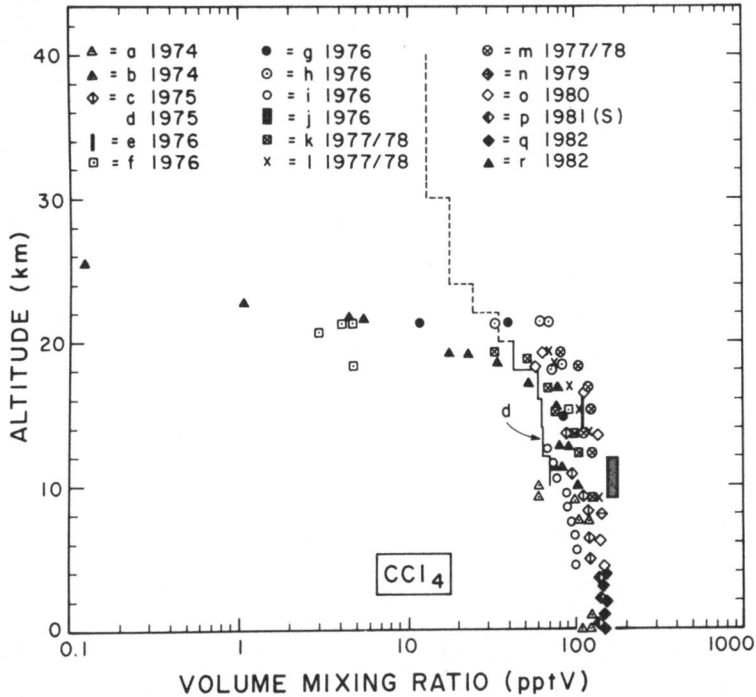

Fig. 11. The vertical distribution of CCl₄ in the atmosphere. Each symbol represents data from a different group of investigators. Unless otherwise marked by (S) behind the year, all data originate from the northern hemisphere. a [43]; b [100]; c [101]; d [58], e [102]; f (60.5°–78° N); g (29.5°–33° N); h (7.0°–16.5° N) [82]; i [81]; j [69]; k (58°–75° N); l (24°–50° N); m (2°–15° N) [103]; n [104]; o [60]; p [61]; q [40]; r [105]

Leifer et al. [60] found elevated CH_3CCl_3 mixing ratios in the Mt. St. Helens eruption plume (c), as compared to other samples at the same height. This is most likely due to the fact that this plume largely consisted of air which originated from lower altitudes and hence showed almost tropospheric abundances of methyl chloroform.

CHF_2Cl (CFC-22)

A preliminary profile of CFC-22, as obtained from the analysis of cryogenically collected balloon samples, indicated CFC-22 mixing ratios decreasing from about 50 pptV at 10 km to about 10 pptV at 33.2 km [90]. This relatively slow fall off in the stratosphere is also indicated by a set of aircraft measurements up to about 20 km carried out by Leifer et al. [60]. It suggests that CFC-22 is indeed fairly stable, thus carrying chlorine well into the higher stratosphere.

The first 2 complete and consistent vertical profiles of CFC-22 were obtained by Fabian et al. [107] by applying a modified cryogenic sampling technique. They are shown in Fig. 13 along with results published by other groups. This data set appears to be consistent except for the high value of 100 pptV, measured at 15 km by Goldman et al. [86] using infrared spectroscopy. The rapid buildup rate of

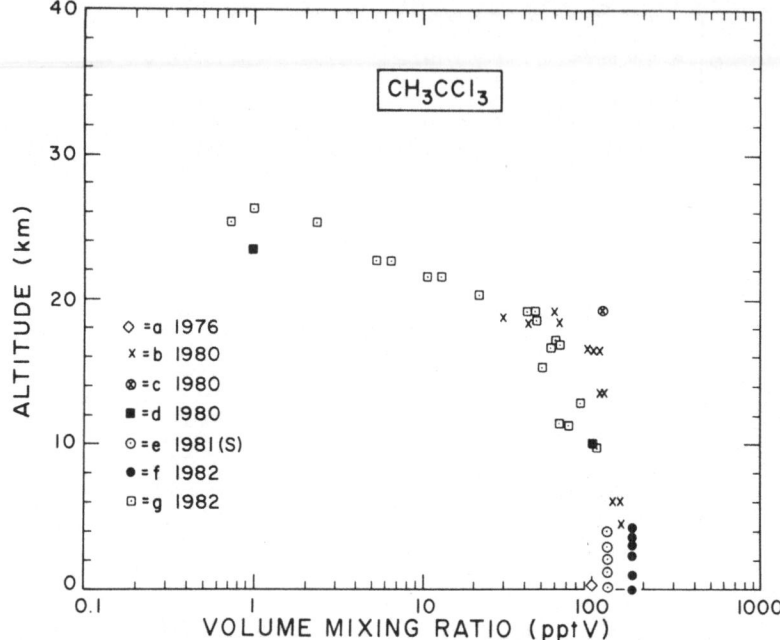

Fig. 12. The vertical distribution of CH_3CCl_3 in the atmosphere. Each symbol represents data from a different group of investigators. Unless otherwise marked by (S) behind the year, all data originate from the northern hemisphere. a [53]; b [60]; c (Mount St. Helens Plume, [60]); d [90]; e [61]; f [40]; g [105]

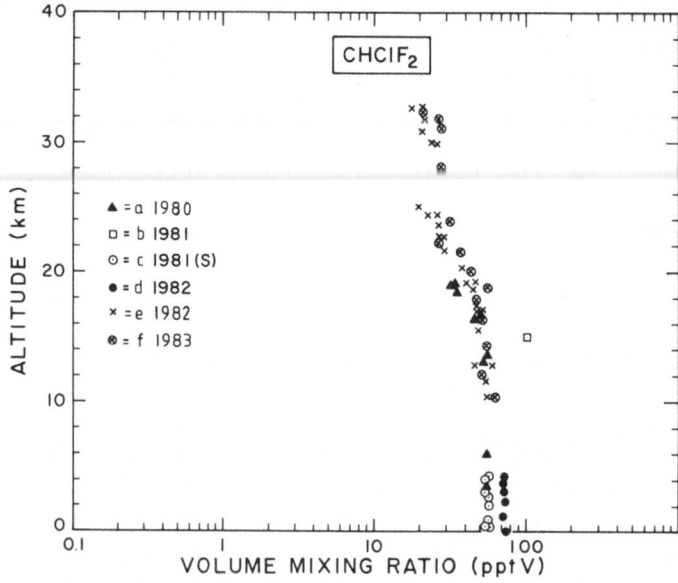

Fig. 13. The vertical distribution of $CHClF_2$ in the atmosphere. Each symbol represents data from a different group of investigators. The time of the measurement is also given in the figure. Unless otherwise marked by (S) behind the year, all data originate from the northern hemisphere. a [60]; b [86]; c [61]; d [40]; e [107], f [107]

CFC-22 in the atmosphere is noticeable from the comparison of the two tropospheric profiles c and d measured 6 months apart in the southern and northern hemispheres, respectively. Model computations show that CFC-22 will soon become the dominating source of odd chlorine above 40 km if the present growth of the global emission remains unabated [107].

Sources, Budgets and Distribution of Other Atmospheric Halocarbons

CF_4 (CFC-14)

The presence of carbon tetrafluoride in the atmosphere was first deduced by Gassmann [108] from MS-analysis of contaminant levels of CF_4 in high-purity krypton samples. Gassmann, suggesting that CF_4 was enriched from the atmosphere during the cryogenic preparation of high-purity krypton, deduced the atmospheric CF_4 abundance possibly ranging between 0.1 and 1 ppbV. Since global industrial production of CF_4, about 10 t/year, was almost negligible, Gassmann suggested that natural exhalation [109] might have built up such large atmospheric CF_4 inventory. Although direct measurements in ambient air by various groups later showed lower mixing ratios than those given by Gassmann – an average of (67 ± 10) pptV may be inferred from several measurement series carried out by Rasmussen et al. [110] and Penkett et al. [111] – the question of the source has remained open.

The composition of samples taken from Mt. Erebus and from Mauna Loa and Kilauea volcanic emissions make a volcanic contribution to the atmospheric CF_4 inventory very unlikely [111]. From elevated CF_4 levels found in plumes of aluminium plants on the Columbia River, Penkett et al. [111] concluded that the aluminium industry is likely to be the major source. According to a study of the Hoechst Company [112], however, the CF_4 production in the electrolytic aluminium reduction furnace is a discontinuous process, restricted to a few minutes per day, as the alumina content of the cell approaches depletion. Thus the global aluminium production between 1950 and 1975 accounts for a total of about 170,000 t of CF_4, building up only 20% of the present abundance.

Carbon tetrafluoride is possibly the most stable fluorocarbon gas. No chemical decomposition reactions are known, and no significant photodissociation can be expected in the lower atmosphere, since CF_4 absorbs only at very short wavelengths below about 100 nm [113]. Cicerone [114] calculated an atmospheric lifetime for CF_4 of more than 10,000 years, and recent measurements of CF_4 absorption cross-sections at Lyman-α [115] confirm that such a long lifetime must be considered realistic. In view of this, even a very inefficient source may have built up the atmospheric CF_4 inventory. Since industrial sources appear to contribute to this inventory, a gradual increase of the CF_4 abundance as predicted by Cicerone [114], must be envisaged.

The presently available measurements show that CF_4 is evenly distributed in the atmosphere. Its average mixing ratio was found to be (67 ± 10) pptV [110–111]. The first vertical profile, measured by Fabian et al. [90] is shown in Fig. 14. It confirms that up to 33 km altitude no significant decomposition occurs. From the quoted precision of these data, given as $\pm 15\%$, it cannot be decided whether or not the slight decrease from about 70 pptV at 10 km to 60 pptV at 33 km is real.

C_2F_6 (CFC-116)

Carbon hexafluoride was, like CFC-14, found to be released from aluminium plants. The ratios of C_2F_6 in the plume to background air, ranging between 2.50 and 2.85, were found to be slightly higher than those for CF_4 [111]. There are as yet not other sources known.

No decomposition processes are reported other than photolysis at wavelengths shorter than about 100 nm [113], suggesting a very long atmospheric lifetime of this molecule, which may range between 500 and 1,000 years.

The first measurements suggest average tropospheric mixing ratios between 3 and 5 pptV [111, 104] with no indication of latitudinal or hemispheric differences. The first vertical profile [90] shows a slight decrease of mixing ratios with altitude, from 4 pptV at 10 km to 2.5 pptV at 33 km (Fig. 14). This difference from CFC-14 may be due to a more rapid buildup in the atmosphere or due to stratospheric decomposition.

$C_2F_3Cl_3$ (CFC-113) and $C_2F_4Cl_2$ (CFC-114)

CFC-113 is used primarily as a solvent, although it can be used as a blowing agent and as a refrigerant; CFC-114 is used chiefly as an aerosol propellant and as a

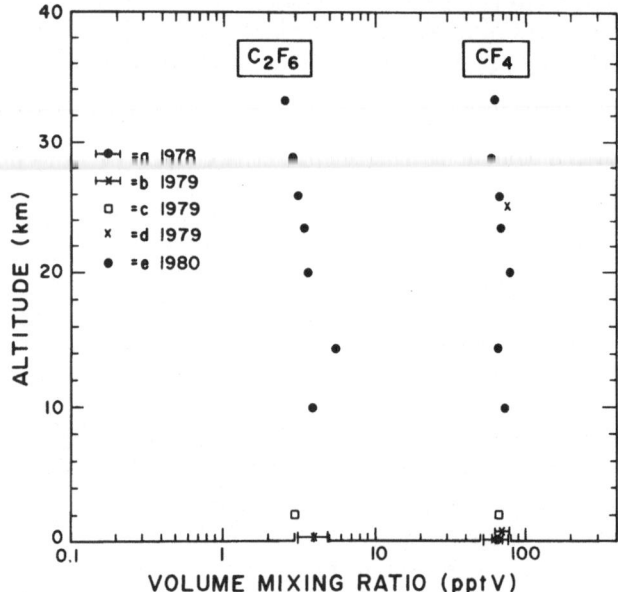

Fig. 14. The vertical distributions of CF_4 and C_2F_6 in the atmosphere. Each symbol represents data from a different group of investigators. The time of measurement is also given in the figure. a [110]; b [111], c [104]; d [85]; e [90]

refrigerant [99], and anthropogenic emissions are likely to be the only sources of these molecules. Present global release rates, as evaluated by Wuebbles [116], are 91×10^3 and 18×10^3 tons per year, for CFC-113 and CFC-114, respectively.

Both constituents were first measured in the atmosphere by Singh et al. [17]. On a latitudinal survey between 60° N and 40° S, during November 1975, they measured average mixing ratios of (19 ± 3.5) pptV and (18 ± 3.1) pptV for CFC-113, and of (12 ± 1.9) pptV and (10 ± 1.3) pptV for CFC-114, for the northern and southern hemispheres, respectively. Although an accurate emission inventory for these species was not available, Singh et al. concluded atmospheric lifetimes of both CFC-113 and CFC-114 likely to be at least several decades, and photolysis in the stratosphere to be the major sink.

A latitudinal survey between 90° N and 90° S, carried out in 1978 by Rasmussen and Khalil [73], yielded average CFC-113 mixing ratios of only (13 ± 0.8) pptV and (12 ± 0.6) pptV, for the northern and southern hemispheres, respectively. The same group measured (20 ± 1) pptV at 20° N in 1980 [117], and (23 ± 1) pptV at 70° N, during May of 1982 [40], while (17 ± 1) pptV were obtained between 30° and 40° S, in November 1981 [61].

With this limited data base only a rough budget calculation can be performed. For CFC-113 an average abundance of 18 pptV would correspond to a global tropospheric burden of 0.6 Mt. Considering the long atmospheric lifetime, which was assessed by Chou et al. [99] to vary between 63 and 122 years, the quoted emission of 91 kt/y would result in a 15% annual increase of the abundance of CFC-113. Such an increase is noticeable from the tropospheric profiles e and f in Fig. 15. For CFC-114, taking an average global abundance of 11 pptV corresponding to an atmospheric burden of 0.3 Mt, the quoted emission of 18 kt/y would result in a 6% annual growth in the atmosphere. The atmospheric lifetime of CFC-114, likely to range between 126 and 310 years [99], is even longer than that of CFC-113.

The first stratospheric profiles of CFC-113 and CFC-114, shown in Fig. 15, were obtained by analyzing cryogenically collected, stratospheric samples from a balloon flight made during September 1980 [90]. It is interesting to note that the profile of CFC-113, containing 3 atoms of fluorine, falls off more rapidly than the profile of CFC-114, which contains 4 fluorine atoms. This confirms the general rule that the stability of fully halogenated hydrocarbons increases with the number of fluorine atoms that replace chlorine atoms (see Fig. 18).

C_2F_5Cl (CFC-115) and CF_3Cl (CFC-13)

CFC-115 is used as a propellant for foods dispensed from aerosols, and as a refrigerant either directly or in mixture with CFC-22 in the azeotropic refrigerant 502; CFC-13 is used as a refrigerant either by itself or in an azeotrope with CHF_3 (refrigerant 503) for very low temperature applications [99]. Their abundance in the atmosphere is certainly due to anthropogenic emission only [111]. The present release rate of CFC-115, as evaluated by Wuebbles [116], is 4.5×10^3 t/y.

Both molecules are very stable in the troposphere, and the only important decomposition process is UV photolysis in the stratosphere. Estimates of the atmo-

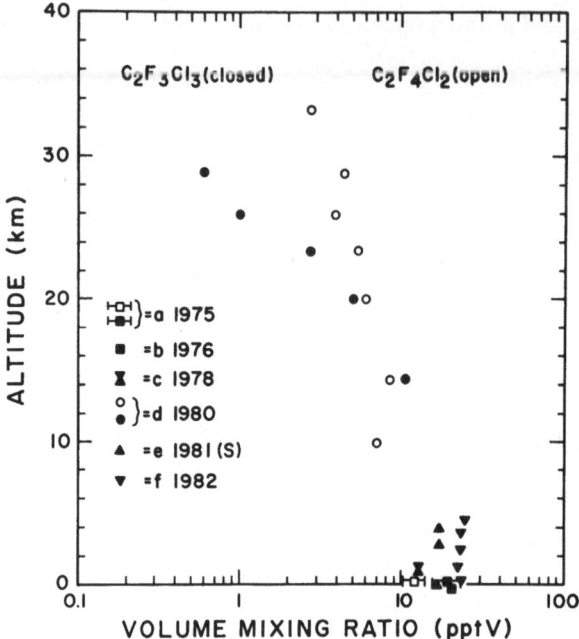

Fig. 15. The vertical distribution of $C_2F_3Cl_3$ and $C_2F_4Cl_2$ in the atmosphere. Each symbol represents data from a different group of investigators. The time of measurements is also given in the figure. Unless otherwise marked by (S) behind the year, all data originate from the northern hemisphere. a [17], b [118]; c [73]; d [90]; e [61]; f [40]

spheric lifetimes range from 230 to 550 years for CFC-115 and from 180 to 450 years for CFC-13 [99].

CFC-13 volume mixing ratios, obtained during 1979 for Antartica and clean northern midlatitude air, are reported as (3.6 ± 0.7) pptV and (3.2 ± 0.5) pptV, respectively [111], and 3 pptV were measured by Rasmussen and Khalil [104] for both hemispheres. The first stratospheric profile [90], as analyzed from cryogoni cally collected balloon samples, suggests tropospheric mixing ratios of the order of 5 pptV (Fig. 16). In the stratosphere, the CFC-13 mixing ratios decrease with height to 2.3 pptV at 33 km. Since no global release rates are assessed yet, it is uncertain whether this stratospheric decrease is due to the buildup in the atmospheric or due to stratospheric decomposition.

Only one measurement of the tropospheric abundance of CFC-115 is published as yet. Penkett et al. [111] give a value of 4.1 pptV, which corresponds to 110,000 tons of C_2F_5Cl that have been released into the atmosphere. The first stratospheric profile measured by Fabian et al. [90] is interesting in comparison with that of CFC-13. It was suggested earlier [111] that CFC-115 is growing more rapidly than other anthropogenic species containing the CF_3 grouping, possibly because it is used in conjunction with CFC-22 in the refrigerant CFC-502. This may account for the steeper profile of CFC-115, whose mixing ratios decrease from 4 pptV at 10 km to about 1 pptV at 30 km. From the quoted emission rates, it follows that the atmospheric abundance of CFC-115 is bound to increase by 5% every year.

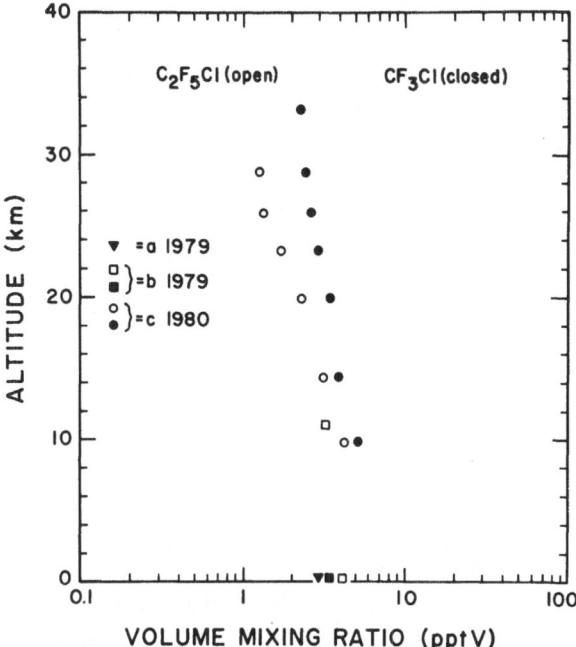

Fig. 16. The vertical distributions of C_2F_5Cl and CF_3Cl in the atmosphere. Each symbol represents data from a different group of investigators. The time of measurement is also given in the figure. a [104]; b [111]; c [90].

Methyl Bromide (CH$_3$Br) and CF$_3$Br (CFC-13B1)

Very limited information is available about these bromine-bearing species. Methyl bromide is, like methyl chloride, most likely of natural origin, while CF_3Br, which is used as a fire extinguisher, has anthropogenic sources.

Singh et al. [17, 53, 118] measured CH_3Br ground-level background mixing ratios of 5 pptV or less, with values as high as 20 pptV in the clean marine environment, indicating that the ocean is likely to be the main source. Rasmussen and Khalil [104] indicated that tropospheric CH_3Br mixing ratios may vary between 5 and 25 pptV, while Penkett et al. [111] found about 10 pptV in the upper troposphere. The reaction with OH controls the removal of CH_3Br from the atmosphere, with an estimated tropospheric lifetime of about 1.7 years [119]. The first stratospheric measurement [90] indicates that CH_3Br falls of rapidly above the tropopause: 1.2 pptV was measured at 14.4 km, while a sample taken at 20 km was already below the detection limit (Fig. 17).

CF_3Br was first measured in the atmosphere by Penkett et al. [111]. Its present abundance of about 1 pptV indicates that the anthropogenic bromine input into the stratosphere is equivalent to about 10% of the natural one. Molina et al. [119] have investigated the photochemistry of CF_3Br. According to this study, it appears that UV photolysis is the main decomposition process in the stratosphere, with long atmospheric lifetimes ranging between 62 and 112 years. Thus a gradual buildup of CF_3Br in the atmosphere can be expected. A first stratospheric profile

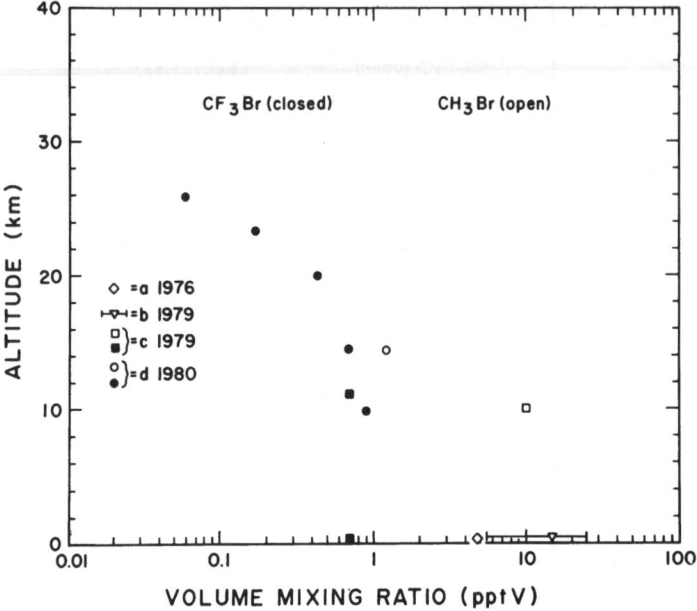

Fig. 17. The vertical distribution of CF_3Br and CH_3Br in the atmosphere. Each symbol represents data from a different group of investigators. The time of measurement is also given in the figure. a [53]; b [104]; c [111]; d [90]

[90] is shown in Fig. 17, with mixing ratios falling off slightly more rapidly than those of the theoretical profile calculated by Molina et al. [119].

Methyl Iodide (CH_3I)

Methyl iodide was first detected in the atmosphere by Lovelock et al. [37], with an average mixing ratio of 1.2 pptV. From the large difference between air and sea concentrations of this species they concluded the ocean to be the major source. Taking an atmospheric lifetime of 50 h, they estimated a global emission rate of 40 Mt/y, methyl iodide thus being the halocarbon with by far the largest emission rate. More recent investigations show that the atmospheric lifetime of CH_3I, with an average of about 5 days [24, 61, 120], is indeed very short. Thus the enormous natural emission rate from the ocean is essentially confirmed.

Worldwide measurements show that CH_3I is almost uniformly distributed over the oceans, with average mixing ratios ranging between 2 and 3 pptV in the planetary boundary layer [61, 73, 117, 121]. There appear to exist regions of high marine biomass productivity, such as upwelling waters off the Peruvian and South African coasts, where averages ranging between 7 and 22 pptV were measured [121]. Above the boundary layer, up to about 2.5 times lower concentrations were found than within the boundary layer [61, 117, 121]. CH_3I mixing ratios measured over the continents appear to be lower than those found over the oceans [17, 40, 121].

Due to its short atmospheric lifetime, CH_3I has certainly no direct influence on the stratospheric halogen budget. As Chameides and Davis have suggested [122], methyl iodide may have some influence on ozone, NO_x and HO_x radicals in the troposphere. More important, however, since CH_3I seems to play a key role in the marine production of CH_3Cl, it is an important precursor substance. In this context it is interesting to note that elevated CH_3I mixing ratios were measured in regions where higher CH_3Cl concentrations were found [121].

Concluding Remarks

Figure 18 shows a summary of the vertical distributions of fully halogenated methanes (a) and ethanes (b) in relative units, with respect to the average tropospheric mixing ratios. Average profiles, derived from the available data as displayed in the Figs. 8–16, are shown. As Chou et al. [99] has pointed out, UV absorption cross-sections depend strongly on the number of chlorine atoms attached to a particular carbon atom. The photon absorption by molecules containing only one Cl atom per molecule is so weak that it is less important as a removal process for such molecules in the stratosphere than reactions with O('D) atoms. Thus, CFC-14 and CFC-116, which contain no chlorine, are the most stable of these substances. CCl_4, with 4 chlorine atoms being less stable than partially fluorinated methanes, is almost entirely photolyzed below 26 km. With 3 chlorine atoms, CFC-113 is the least stable of the ethanes discussed in this review. It is evident that the whole set of halocarbon profiles provides a powerful tool for testing the UV radiation flux schemes of atmospheric models to resolve presently existing discrepancies.

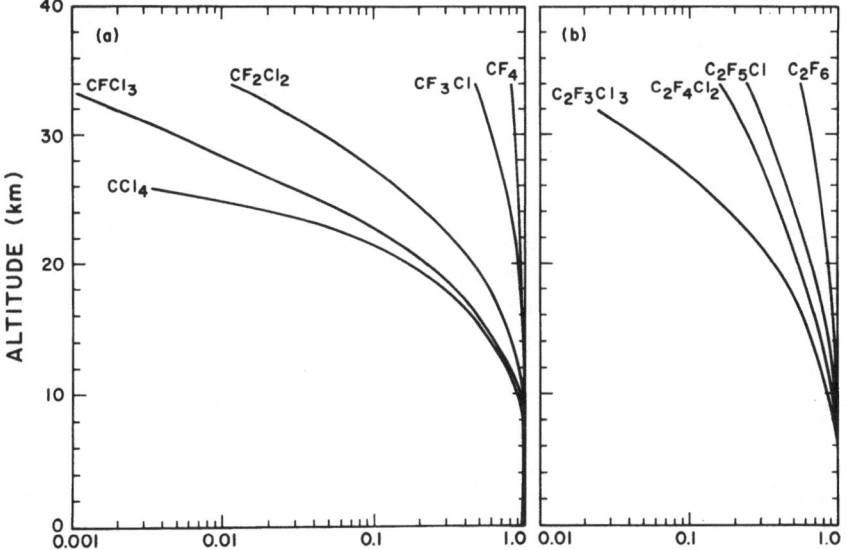

Fig. 18. Averaged vertical profiles of fully halogenated methanes (a) and ethanes (b) in relative units with respect to the tropospheric abundance

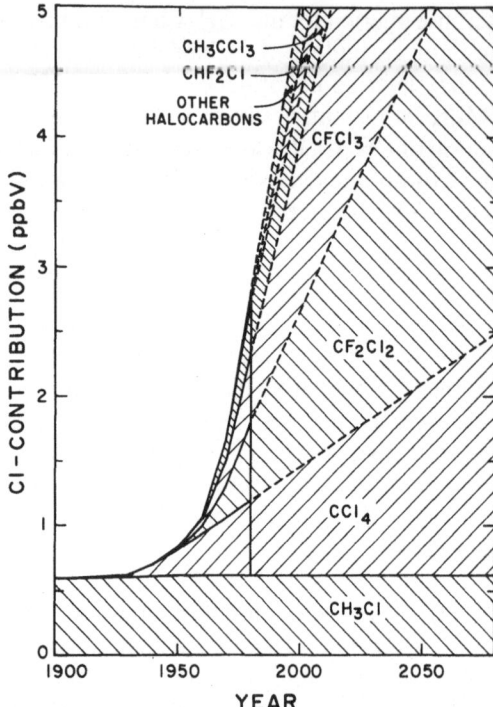

Fig. 19. Sources of organically bound chlorine in the atmosphere as a function of time. The solid curves until 1980 base upon the evaluated emission rates. The projections for future years (dashed lines) base upon constant emissions at the 1980 level.

Since the different halocarbons are decomposed at different altitudes, the chlorine release from these source gases is accordingly height-dependent. Wuebbles and Chang [123] have calculated the steady-state production rates of atomic chlorine for the individual halocarbons as a function of altitude. From this study, it follows for instance that the chlorine production from CFC-11 is largest slightly below 30 km, whereas that from CFC-115 has a maximum around 40 km.

The data base compiled in this review enables the assessment of the past and present source of organically bound chlorine and its projection for future years. Figure 19 shows the result of this assessment indicating almost 3 ppbV (2.86 ppbV) of organically bound chlorine in the atmosphere by 1980, 78% of which are of anthropogenic origin. Only the long-lived species are included in this figure, short-lived substances such as CH_2Cl_2, $CHCl_3$, $CCl_2 = CCl_2$, and $CHCl = CCl_2$ were not considered.

For the future projection, given by the dashed lines, it was assumed that all emission rates remain constant at the 1980 level. This is probably a conservative estimate, since at least the global emission rates of CH_3CCl_3, CFC-22, and the "other halocarbons", i.e., CFC-113, CFC-114, CFC-115, and CFC-13, are likely to increase during the coming years. This may be partly counteracted by the fact that some of the halocarbons approach steadystate conditions. Thus, it appears likely that the abundance of organically bound chlorine will exceed 5 ppbV by the end of this century, 88% of which then being of anthropogenic origin.

Acknowledgements

This article was prepared during a stay at the Chemistry Department of the University of California at Irvine. The author is indebted to the Max-Kade-Foundation for granting a research fellowship, which made this stay possible. Helpful discussions with Drs. F. S. Rowland of UCI, Y. L. Yung of CalTech, M. J. Molina of JPL, D. J. Wuebbles of Lawrence Livermore Lab. and R. C. Whitten and J. F. Vedder of the NASA Ames Research Center are gratefully acknowledged.

References

1. Stolarski R. S., Cicerone, R. J.: Can. J. Chem. 52, 1610 (1974)
2. Wofsy, S. C., McElroy, M. B.: Can. J. Chem. 52, 1582 (1974)
3. Crutzen, P. J.: Can. J. Chem. 52, 1569 (1974)
4. Cicerone, R. J., Stolarski, R. S., Walters, S.: Science 195, 1165 (1974)
5. Molina, M. J., Rowland, F. S.: Nature 249, 810 (1974)
6. Rowland, F. S., Molina, M. J.: Rev. Geophys. Space Phys. 13, 1 (1975)
7. NASA Ref. Publ. 1010: Chlorofluoromethanes and the stratosphere (R. D. Hudson, Ed.), Goddard Space Flight Center, 1977
8. Logan, J. A., Prather, M. J. Wofsy, S. C., McElroy, M. B.: Trans. Roy. Soc. (London) 290, 187 (1978)
9. Crutzen, P. J., Isaksen, I. S. A., McAffee, J. R.: J. Geophys. Res. 83, 345 (1978)
10. NASA Ref. Publ. 1049: The Stratosphere: Present and Future (R. D. Hudson and E. I. Reed, Eds.), Goddard Space Flight Center, 1979
11. NATO Advanced Study Institute on Atmospheric Ozone: Its Variation and Human Influences (M. Nicolet and A. C. Aikin, Eds.). Report No. FAA-EE-80-20, U.S. Department of Transportation, Washington, D.C., 1980
12. WMO-Report No. 11, The Stratosphere 1981: Theory and Measurements, Geneva, 1981
13. Russow, J.: Fluorocarbons, in: The Handbook of Environmental Chemistry (ed.) O. Hutzinger, 3A, p. 133. Berlin-Heidelberg-New York, Springer 1980
14. Pearson, C. R.: C_1 and C_2 Halocarbons, in: The Handbook of Environmental Chemistry (ed.). O. Hutzinger, 3B, p. 69. Berlin-Heidelberg-New York, Springer 1982
15. Lovelock, J. E.: Nature 256, 193 (1975)
16. Grimsrud, E. P., Rasmussen, R. A.: Atmos. Environ. 9, 1010 (1975)
17. Singh, H. B., et al.: Science 203, 899 (1979)
18. Rasmussen, R. A., et al.: J. Geophys, Res. 85, 7350 (1980)
19. Cicerone, R. J., Stedman, D. H., Stolarski, R. S.: Geophys Res. Lett. 2, 219 (1975)
20. Yung, Y. L., McElroy, M. B., Wofsy, S. C.: Geophys Res. Lett. 2, 397 (1975)
21. Graedel, T. E.: J. Geophys. Res. 84, 273 (1979)
22. Zafiriou, O. C.: J. Mar. Res. 33, 75 (1975)
23. Watson, A. J., Lovelock, J. E., Stedman, D. H.: The Problem of Atmospheric Methyl Chloride, in [11], 365 (1980)
24. Cicerone, R. J.: Rev. Geophys. Space Phys. 19, 123 (1981)
25. Rasmussen, R. A.: Methyl Chloride in the Air of Kenya, final report, Manufacturing Chem. Assoc., 1977, quoted in [18]
26. Crutzen, P. J., et al.: Nature 282, 253 (1979)
27. Inn, E. C. Y., Vedder, J. F., Condon, E. P., O'Hara, D.: Science 211, 821 (1981)
28. Cronn, D. R., Nutmagul, W.: J. Geophys. Res. 87, 11153 (1982)
29. Cowan, M. I., et al.: Trans. Br. Mycol. Soc. 60, 347 (1973)
30. Logan, J. A., et al.: J. Geophys. Res. 86, 7210 (1981)
31. Khalil, M. A. K., Rasmussen, R. A.: Chemosphere 10, 1019 (1981)

32. Stoiber, R. E., et al.: Geol. Soc. Amer. Bull. *82*, 2299 (1971)
33. Chemical Manufacturer's Association (CMA): World Production and Release of CFC-11 and CFC-12, Washington, D.C. 1981
34. CMA, Update of Global Production and Release of CFC-11 and CFC-12, personal communication by Hoechst Aktiengesellschaft, Frankfurt/M. (1982)
35. Jesson, J. P.: Release of Industrial Halocarbons and Tropospheric Budget, in: [11], 373 (1980)
36. Galbally, I. E.: Science *193*, 573 (1976)
37. Lovelock, J. E., Maggs, R. J., Wade, R. J.: Nature *241*, 194 (1973)
38. Rasmussen, R. A., Khalil, M. A. K., Dalluge, R. W.: Science *211*, 285 (1981)
39. Khalil, M. A. K., Rasmussen, R. A.: J. Air Poll. Control Assoc. *31*, 1274 (1981)
40. Rasmussen, R. A. Khalil, M. A. K.: Natural and Anthropogenic Trace Gases in the Lower Troposphere of the Arctic. Dept. of Environmental Science, Oregon Graduate Center, Beaverton, manuscript (1983)
41. Lovelock, J. E.: Nature *230*, 379 (1971)
42. Lovelock, J. E.: Atmos. Environ. *6*, 917 (1972)
43. Lovelock, J. E.: Nature *252*, 292 (1974)
44. Pack, D. H., Lovelock, J. E., Cotton, G., Curthoys, C.: Atmos. Environ. *11*, 329 (1977)
45. Wilkniss, P. R., et al.: Nature *245*, 45 (1973)
46. Wilkniss, P. R., et al.: J. Atmos. Sci. *32*, 158 (1975)
47. Wilkniss, P. R., et al.: Science *187*, 832 (1975)
48. Wilkniss, P. R., et al.: J. Geophys. Res. *83*, 3672 (1978)
49. Zafonte, L. N., Hester, E., Stephens, E. R., Taylor, O. C.: Environ. *9*, 1007 (1975)
50. Hester, N. E., Stephens, E. R., Taylor, O. C.: Environ. Sci. Tech. *9*, 875 (1975)
51. Heidt, L. E., Lueb, R., Pollock, W., Ehhalt, D. H.: Geophys. Res. Lett. *2*, 445 (1975)
52. Singh, H. B., Salas, L. J., Cavanagh, L. A.: J. Air Pollut. Contr. Assoc. *27*, 332 (1977)
53. Singh, H. B., Salas, L. J., Shigeishi, H., Crawford, A.: Atmos. Environ. *11*, 819 (1977)
54. Robinson, E. R., et al.: Geophys. Res. Lett. *3*, 323 (1976)
55. Cronn, D. R., Rasmussen, R. A., Robinson, E., Harsch, D. E.: J. Geophys. Res. *82*, 5935 (1977)
56. Pieretti, D., Rasmussen L. E., Rasmussen, R. A.: Geophys. Res. Lett. *5*, 1001 (1978)
57. Fraser, P. J. B., Pearman, G. I.: Atmos. Environ. *12*, 839 (1978)
58. Krey, P. W., Lagomarsino, R. J., Toonkel, L. E.: J. Geophys. Res. *82*, 1753 (1977)
59. Goldan, P., Custer, W. C., Albritton, D. L., Schmeltekopf, A. L.: J. Geophys. Res. *85*, 413 (1980)
60. Leifer, R., Sommers, K., Guggenheim, S. F.: Geophys. Res. Lett. *8*, 1079 (1981)
61. Rasmussen, R. A., Khalil, M. A. K., Crawford, A. J., Fraser, P. J.: Geophys. Res. Lett. *9*, 701 (1982)
62. Penkett, S. A., et al.: Atmos. Environ. *13*, 1011 (1979)
63. Brice, K. A., et al.: Atmos. Environ. *16*, 2543 (1982)
64. Bortoli, M. de, Pecchio, E.: Atmos. Environ. *10*, 921 (1976)
65. Jaffar, M., Dutkiewicz, V. A., Husain, L.: Atmos. Environ. *15*, 1653 (1981)
66. Singh, H. B., Fowler, D. P., Peyton, T. O.: Science *192*, 1231 (1976)
67. Graedel, T. E., Allara, D. L.: Atmos. Environ, *10*, 385 (1976)
68. Simmonds, P. G., et al.: J. Geophys. Res. *88*, 8427 (1983)
69. Tyson, B. J., Arvesen, J. C., O'Hara, D.: Geophys. Res. Lett. *5*, 535 (1978)
70. Neely, W. B., Plonka, J. H.: Environ. Sci. Techn. *12*, 317 (1978)
71. Makide, Y., Rowland, F. S.: Proc. Natl. Acad., Sci. USA *78*, 5933 (1981)
72. Rasmussen, R. A., Khalil, M. A. K.: Geophys. Res. Lett. *8*, 1005 (1981)
73. Rasmussen, R. A., Khalil, M. A. K.: Chemosphere *11*, 227 (1982)
74. Lovelock, J. E.: Nature. *267*, 32 (1977)
75. Rowland, F. S., Tyler, S. C., Montague, D. C.: Global Distributions of CH_3CCl_3, CCl_3F, and CCl_2F_2 in July 1979, in (J. London, ed.) Proc. Int. Ozone Symp. Boulder, Colorado (1981)
76. Khalil, M. A. K.: Topics in the Behavior of Atmospheric Trace Gases, Ph.D. dissertation, Oregon Graduate Center, Beaverton, OR, 1979
77. Rasmussen, R. A., Khalil, M. A. K., Penkett, S. A., Prosser, N. J. D.: Geophys. Res. Lett. *7*, 809 (1980)

78. Khalil, M.A.K., Rasmussen, R.A.: Nature *292*, 823 (1981)
79. NASA Panel for Data Evaluation: Chemical Kinetics and Photochemical Data for Use in Stratospheric Modeling. JPL Publication 82–57, Pasadena, California (1982)
80. Schmeltekopf, A.L., et al.: Geophys. Res. Lett. *2*, 393 (1975)
81. Seiler, W., Müller, F., Oeser, H.: Pure Appl. Geophys. *116*, 554 (1978)
82. Vedder, J.F., et al.: Geophys. Res. Lett. *5*, 33 (1978)
83. Fabian, P., et al.: J. Geophys. Res. *84*, 3149 (1979)
84. Fabian, P.: Adv. Space Res. *1*, No. 11, 17 (1981)
85. Goldman, A., et al.: Geophys. Res. Lett. *6*, 609 (1979)
86. Goldman, A., et al.: Geophys. Res. Lett. *8*, 1012 (1981)
87. Robinson, E.R., et al.: Atmos. Environ. *11*, 213 (1977)
88. Penkett, S.A., Derwent, R.G., Fabian, P., Borchers, R., Schmidt, U.: Nature *283*, 58 (1980)
89. Schmidt, U., et al.: The Vertical Distribution of CH_3Cl, $CFCl_3$ and CF_2Cl_2 in the Midlatitude Stratosphere, in (J. London, ed.) Proc. Int. Ozone Symp., Boulder, Colorado (1981)
90. Fabian, P., Borchers R., Penkett, S.A., Prosser, N.J.D.: Nature *294*, 733 (1981)
91. Krey, P.W., Lagomarsino, R.J., Frey, J.J.: J Geophys. Res. *81*, 1557 (1976)
92. Vedder, J.F., Inn, E.C.Y., Tyson, B.J., Boitnott, A.: J. Geophys. Res. *86*, 7363 (1981)
93. Fabian, P., et al.: J. Geophys. Res. *86*, 5179 (1981)
94. Froidevaux, L., Yung, Y.L.: Geophys. Res. Lett. *9*, 854 (1982)
95. Frederick, J.E., Mentall, J.E.: Geophys. Res. Lett. *9*, 461 (1982)
96. Herman, J.R., Mentall, J.E.: J. Geophys. Res. *87*, 8967 (1982)
97. Hunten, D.M.: Geophys Res. Lett. *10*, 333 (1983)
98. Chou, C.C., et al.: J. Phys. Chem. *81*, 286 (1977)
99. Chou, C.C., et al.: J. Phys. Chem. *82*, 1 (1978)
100. Cox, R.A., Derwent, R.B., Eggleton, A.E.J.: Atmos. Environ. *10*, 305 (1976)
101. Robinson, E., et al.: Atmos. Environ. *11*, 215 (1977)
102. Leifer, R., Larsen, R., Toonkel, L.: Stratospheric Distributions and Inventories of Trace Gases in the Northern Hemisphere for 1976, in: Report EML-349, I-2111, U.S. Depart. of Energy (1973)
103. Leifer, R., Toonkel, L., Larsen, R.: Project Airstream, Trace Gases in the Stratosphere, in: Report EML-349, II-107, U.S. Dept. of Energy (1979)
104. Rasmussen, R.A., Khalil, M.A.K.: Atmospheric Halocarbons, Measurements and Analyses of Selected Trace gases, in (11), 209 (1980)
105. Borchers, R., Fabian, P., Penkett, S.A.: First Measurements of the Vertical Distribution of CCl_4 and CH_3CCl_3 in the Stratosphere. Die Naturwiss. *70*, 514 (1983)
106. Gömer, D.: Simulation von Spurenstoffverteilungen in der Atmosphäre mit Hilfe von ein- und zweidimensionalen Modellrechnungen, Ph. D. Dissertation, University of Göttingen, 1983
107. Fabian, P., Borchers R., Krüger, B.C., Lal, S., Penkett, S.A.: Geophys. Res. Lett. *12*, 1 (1985)
108. Gassmann, M.: Die Naturwiss. *61*, 127 (1974)
109. Kranz, R.: Die Naturwiss. *53*, 593 (1966)
110. Rasmussen, R.A, Penkett, S.A., Prosser, N.: Nature *277*, 549 (1979)
111. Penkett, S.A., Prosser, N.J.D., Rasmussen, R.A., Khalil, M.A.K.: J. Geophys. Res. *86*, 5172 (1981)
112. Hoechst Aktiengesellschaft, internal study 1976, O. Klais, personal communication (1982)
113. Sandorfy, C.: Atmos. Environ. *10*, 343 (1976)
114. Cicerone, R.J.: Science *206*, 59 (1979)
115. Inn, E.C.Y.: J. Geophys. Res. *85*, 7493 (1980)
116. Wuebbles, D.J.: J. Geophys Res. *88*, 1433 (1983)
117. Rasmussen, R.A., Khalil, M.A.K.: Pure Appl. Geophys. *119*, 990 (1981)
118. Singh, H.B.: Geophys. Res. Lett. *4*, 101 (1977)
119. Molina, L.T., Molina, M.J., Rowland, F.S.: J. Phys. Chem. *86*, 2672 (1982)
120. Zafiriou, O.C.: J. Geophys. Res. *79*, 2730 (1974)
121. Rasmussen, R.A., et al.: J. Geophys. Res. *87*, 3086 (1982)
122. Chameides, W.L., Davis, D.D.: J. Geophys. Res. *85*, 7383 (1980)
123. Wuebbles, D.J., Chang, J.S.: J. Geophys. Res. *86*, 9869 (1981)

Formation, Transport and Control of Photochemical Smog

Hans Güsten

Kernforschungszentrum Karlsruhe, Institut für Radiochemie
D-7500 Karlsruhe, Federal Republic of Germany

Summary

The present knowledge of the chemistry of the reactions involved in photochemical air pollution is discussed in terms of the elementary photochemical and chemical reactions. The importance of the necessary precursors NO_x and hydrocarbons on the maximum ozone levels in polluted urban air as well as the nature of the free radical chain mechanisms and the potential of production of reactive radicals in catalytic NO to NO_2 conversion are outlined. Since the OH radical is recognized as the primary attacking species initiating the oxidation and degradation of organic and inorganic compounds, the OH rate coefficient is used to assess the lifetime of organic compounds in the troposphere.

After a general discussion of the laboratory experiments performed in smog chambers and of the mathematical simulation of the photosmog chemistry in computer modelling studies, a summary is presented of the formation, occurrence and transport of photochemical air pollution in Europe. In particular, the concentrations of ozone, PAN, hydrogen peroxide, the nitrate radical, nitric and nitrous acid, and aldehydes and ketones are reviewed. The phenomenon of long-range transport of photochemical oxidants over distances of 1,000 km adjacent to source regions and its mathematical description in transport models are considered. Finally, control strategies are discussed of ozone on the basis of air quality criteria and standards, on the isopleths of the precursor concentrations of photochemical air pollution as well as on societal options.

Introduction

Air pollution was originally regarded as a sign of industrial activity and economic progress. The air resource of the earth was considered as limitless. However, the exponentially growing population and industrializing society have caused serious air pollution problems on a local and global scale during the second half of the twentieth century. In particular, local air resources are often restricted by geographical and meteorological effects. The classical Los Angeles type smog, generally called photochemical smog, has received increasing attention in Europe during the last decade. Until about 1970 the phenomenon of photochemical smog was considered as a specific air pollution problem of the Los Angeles metropolitan area. Thus, it is not surprising that in the 1950s and 1960s nearly all the scientific work on photochemical smog formation in field investigations and laboratory experiments was carried out in the US. The results of this research have been summarized in a number of reviews [1–5]. In the early 1970s, photochemical air pollution episodes similar to those in Los Angeles were observed in many large metropolitan areas throughout the world. This article focuses on the formation, occurrence and transport of photochemical air pollution with special emphasis on the situation in Europe.

Historical Background

In the late 1940s and early 1950s a new phenomenon was observed in the Los Angeles area. On sunny days with anticyclonic weather conditions when the automobile traffic level was high, eye irritation was a common complaint of the citizen. Analyses of the air by Haagen-Smit et al. [6, 7] showed that a high level of oxidants, mainly ozone, was formed by gas-phase oxidation reactions. Gross photochemical oxidants are primarily ozone but include also other species, for example NO_2 and PAN, capable of oxidizing aqueous solutions of iodide ions. Other symptoms of photochemical smog are the reduction of the visibility and typical plant damage. It was soon learned in laboratory studies carried out in so-called smog chambers that the action of simulated sunlight on dilute mixtures of automobile exhaust in air resulted in the same symptoms such as ozone formation

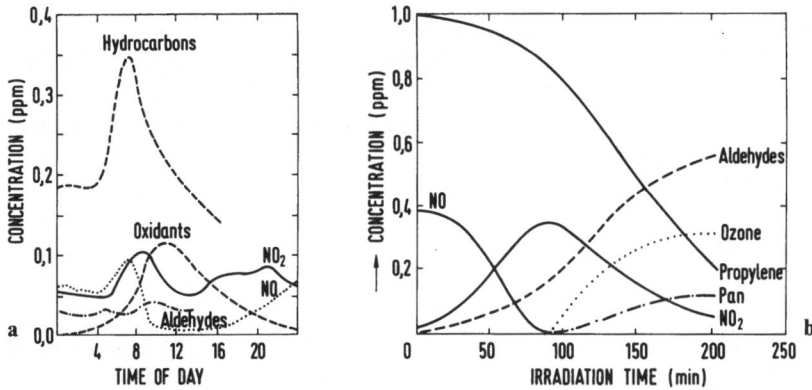

Fig. 1. a) The diurnal variation of photochemical smog buildup in Los Angeles [1], b) Concentration-time profiles of various compounds in a UV irradiated mixture of nitric oxide (with some nitrogen dioxide) and propylene in air

[7], eye irritation [8], aerosol formation, wich is responsible for the reduction of visibility [9], and plant damage [10, 11] just as in the free atmosphere.

In Fig. 1 the diurnal variation of the concentration of the oxides of nitrogen and ozone on a smoggy day in Los Angeles are juxtaposed to the time history of the photochemical reaction of a dilute automobile exhaust with light in the wavelength range of 300–420 nm in a smog chamber. In general, the following events take place:
- Nitric oxide (NO) is converted to nitrogen dioxide (NO_2).
- When all of the NO has disappeared, ozone starts to appear.
- The hydrocarbons are oxidized and disappear.
- Reaction products such as aldehydes and PAN are formed.

The primary pollutants, which are those pollutants emitted directly into the atmosphere (the oxides of nitrogen, NO_x, and hydrocarbons), are transformed to secondary pollutants (ozone, aldehydes, PAN, etc.) by chemical and photochemical reactions of the primary pollutants. A detailed review of the formation, reactions and properties of peroxyacetyl nitrates (PAN) in photochemical air pollution was published by Stephens [13] in 1969. Since the early 1970s, photochemical air pollution episodes have been observed in Europe, Japan [14] and Australia [15]. Although photo-oxidants are formed mainly in highly industrial and urban areas, photochemical air pollution has been observed over distances of 1,000 km adjacent to source regions in North-Western Europe [16]. Thus, long-range transport of oxidants raises the question of the large-scale pollution of ecological systems. In particular, the phytotoxicity of photochemical oxidants causes damage to vegetation and crops. It has been established that in the USA ozone could be responsible for over 3×10^9 each year in terms of damage to crops.

National air quality standards for oxidants were first established in the US in 1971 [17] and international guidelines for ozone were proposed by a working group of the World Health Organization [18].

Since the beginning of the last decade, extensive photochemical kinetic mechanisms together with computer simulation of the chemistry of the urban atmosphere were developed. A major goal of the computer simulation is the theoretical prediction of the time history of the photochemical smog process, i.e. the reactant and product concentration versus time curves for proposed kinetic mechanisms. In the distant future these computer simulations might lead to cost-effective control programmes and abatement strategies.

Photochemical Smog Chemistry

The oxides of nitrogen, NO and NO_2, commonly denoted NO_x, play a key role in air pollution chemistry. The principal source of NO_x in the urban atmosphere is the combustion process of fossil fuels, thus, traffic, industry, power plants, residential heating systems, etc. are the principal sources of emission. NO is the dominant species, generally comprising over 95% of the total NO_x. NO_2 is formed from NO in hot combustion exhaust gases by a termolecular reaction:

$$2NO + O_2 \rightarrow 2NO_2.$$

The presence of even small amounts of NO_2 in the atmosphere is sufficient to trigger the complex series of reactions producing photochemical smog. Photochemial smog is formed in the lower troposphere through a series of simultaneous and consecutive chemical reactions which are triggered through the photolysis of NO_2 by solar light in the 300–420 nm wavelength range. At low elevations within the atmosphere there is a cutoff of the solar light at wavelengths of about 290 nm. The energy-rich portion of the sunlight from the ultraviolet limit near 290 nm to wavelengths of about 420 nm causes bond rupture in some sunlight-absorbing atmospheric pollutants. The primary pollutant NO_2, a brown gas, absorbs solar light most efficiently (Fig. 2). The primary quantum yield for the NO_2 photolysis

$$NO_2 + h\nu(\lambda < 420 \text{ nm}) \rightarrow NO + O\,[^3P] \tag{1}$$

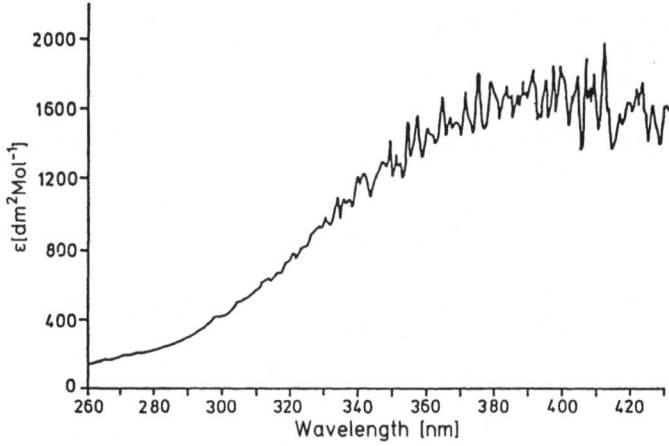

Fig. 2. UV-absorption spectrum of NO_2 below 420 nm [1], ε = decadic extinction coefficient

is close to unity at 366 nm and at shorter wavelengths, i.e. every absorbed solar light quantum leads to the formation of NO and an oxygen atom in its ground state, O [³P]. As the wavelengths of the light is increased to 398 nm the quantum yield for Eq. 1 slightly decreases with a sharp cutoff for wavelengths beyond 400 nm [3, 19–21]. To calculate the specific rate of Eq. 1 the expression

$$J_{NO_2} = \int_\lambda \sigma_\lambda \cdot \Phi_\lambda \cdot I_\lambda d_\lambda \; [\text{time}^{-1}] \tag{2}$$

is used, where σ_λ is the NO_2 absorption coefficient as a function of the wavelengths (Fig. 2), Φ_λ the quantum yields of Eq. 1 as a function of wavelengths, and I_λ the solar actinic flux. Unlike thermal chemical reactions whose reaction rate constants depend on temperature only, the photodissociation coefficient J_{NO_2} depends on the three quantities in Eq. 2. Since I_λ is a time-dependent function determined by the position of the sun, J_{NO_2} with the dimension [time^{-1}] depends on the geographic location, on the season as well as on the local standard time.

Although the photodissociation coefficient J_{NO_2} can be calculated [1], its direct experimental determination by a continuous actinometric method in the troposphere [22–26] and in the stratosphere [27] has been developed recently. Contrary to the calculation of J_{NO_2}, the experimentally determined values account for the fact that the solar irradiation is attenuated within the polluted air layer at various times of observations and latitudes. Also the albedo background for a given location with its particular surface can be taken into account. An updated article on the measurement of the photodissociation coefficient of NO_2 in the atmosphere has been published by Parrish et al. [28].

Basic Photochemical Cycle of NO, NO₂, and Ozone

The rate of photodissociation of NO_2 in Eq. 1 is very fast. In full sunlight at a solar zenith angle of 40°, half of a given NO_2 concentration is depleted in less than two minutes. This reaction is sufficiently fast to trigger the chemistry of the polluted atmosphere. The photolysis of NO_2 to produce O[³P] atoms, followed by their reaction with molecular oxygen, is the primary mechanism for the production of ozone in the lower troposphere.

$$O[^3P] + O_2(+M) \; \rightarrow \; O_3(+M) \; (M = N_2 \text{ or } O_2). \tag{3}$$

M is a chaperone gas, i.e. primarily atmospheric nitrogen or oxygen. The species M carries away excess energy whilst remaining chemically unchanged. The chemical mechanism of ozone generation in polluted air appears to be well understood [1–5, 19, 29, 30]. The formation of ozone has been widely used as a measure of smog intensity. This will be discussed in detail in a forthcoming section. A fraction of the NO emitted in the air parcel reacts with ozone to regenerate NO_2:

$$O_3 + NO \; \rightarrow \; NO_2 + O_2. \tag{4}$$

This reaction is often called "the titration reaction." Since Eq. 4 proceeds very rapidly, a photostationary state between NO, NO_2, and O_3 is set up in which the coexistent concentrations of these three species in the sunlight atmosphere is defined by:

$$\frac{k_1}{k_3} \cong \frac{[NO] \, [O_3]}{[NO_2]} \, , \tag{5}$$

where k_1 and k_3 are the rate constants of Eqs. 1 and 4. Since Eq. 1 is not a thermal reaction, Eq. 5 is better considered as quasi-steady-state. During the day, the fraction of NO converted to NO_2 is often very well approximated by the photostationary state of Eq. 5. If either O_3 or NO is present in ambient air at a high level, the other must necessarily be present at a rather low level. The photostationary state occurs when there is a balance between the photolysis of NO_2 and the oxidation of NO by O_3. Since Eq. 1 is a sunlight-induced reaction, the photostationary state relation is, of course, not valid at night. At night the available NO will react with ozone until either NO or ozone is depleted. In urban areas with continuing NO emissions at night, due, for example, to residential heating or traffic, the night-time levels of ozone are close to zero. A substantial amount of NO_2 generated at night by the reaction of Eq. 4 will participate in the next day's photosmog chemistry. A typical urban diurnal cycle of NO, NO_2 and ozone is shown in Fig. 3.

The morning rush hour adds more NO and reactive hydrocarbons to the atmospheric mixture. In the early morning NO is converted rapidly to NO_2. As a consequence of the photochemical reaction of Eq. 1, peak values of O_3 are reached in the early afternoon. The time lag between maximum precursor emissions of NO_x and maximum ozone concentrations is highly variable and depends

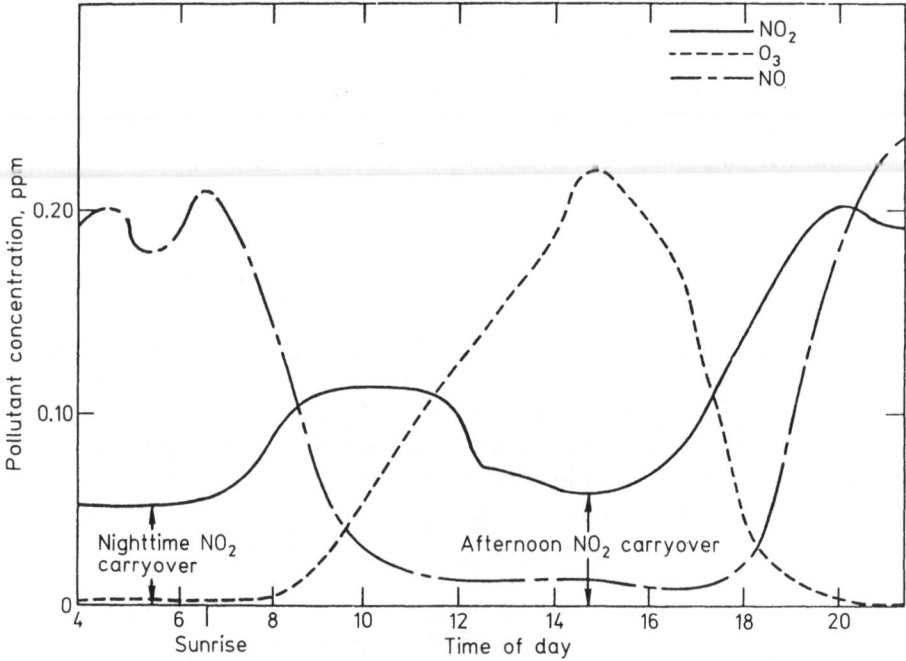

Fig. 3. Urban diurnal cycle of NO, NO_2 and ozone in St. Louis, October 1, 1976 [256]

on the atmospheric turbulence and other meteorological parameters. For this reason maximum O_3 levels are often observed downwind of the urban centre.

The simple relation of Eq. 5 enables a rapid and convenient prediction of ozone levels anticipated in a sunlight-irradiated air mass. The ratios of k_1/k_3 can be calculated in theory and compared with the measured values of the NO, NO_2 and ozone concentrations. Using ambient air data from Los Angeles, the photostationary state relation of Eq. 5 has been tested to be valid within the limits of the reaction schemes of Eqs. 1–5, at least at times after approximately 9:30 a.m. [19]. Before this time and under low light intensities and low concentrations of the reactants, large deviations from the simple theory are observed.

Catalytic Cycle of NO→NO$_2$ Conversion

The "titration reaction" of Eq. 4 should not lead to a steady production of ozone over the day, as is usually measured in urban air (see Fig. 3), since one molecule of O_3 is removed for each NO_2 produced. In order to get a net production of ozone, it is necessary that NO_2 is formed from NO by a reaction other than the O_3-consuming "titration reaction". Any reaction other than that of Eq. 4 which converts NO to NO_2 leads to a net production of ozone. Since the thermal oxidation of NO by molecular oxygen is much too slow due to the very low levels of NO in the well-mixed ambient troposphere, other reactions or catalytically active species have to be responsible for the atmospheric oxidation of NO in photochemical smog. In 1961, Leighton [1] in his classical book on the chemistry of air pollution, suggested that free radicals, predominantly the hydroperoxy (HO_2) and the alkylperoxy (RO_2) radicals involved in photochemical smog chemistry, are responsible for the NO to NO_2 conversion.

In polluted air a great variety of highly reactive transient radicals are formed through both thermal and photochemical reactions. Many kinetic, mechanistic and theoretical studies on the chemistry of a series of reactive radicals have shown that the hydroperoxy (HO_2), alkylperoxy (RO_2) and the acylperoxy radicals ($RCOO_2$) are the dominant species which are the chain carriers in the NO to NO_2 conversion.

$$(RO_2, HO_2) + NO \rightarrow (RO, OH) + NO_2$$
$$RCOO_2 + NO \rightarrow RCO_2 + NO_2. \tag{6}$$

In these catalytic cycles NO is not used up. The amount of ozone produced is now greater than the amount of NO_x present. The OH radicals and alkyloxy radicals (RO) produced in Eq. 6 can react with hydrocarbons, aldehydes or carbon monoxide (CO) present in the polluted air in which they efficiently generate HO_2 and RO_2 radicals.

$$OH + RH \rightarrow R + H_2O \tag{7}$$

$$R + O_2 \rightarrow RO_2 \tag{8}$$

$$OH + CO \rightarrow CO_2 + H \tag{9}$$

$$H + O_2(+M) \rightarrow HO_2(+M) \tag{10}$$

$$RO + O_2 \rightarrow HO_2 + R'CHO \tag{11}$$
$$\text{(aldehyde or ketone)}$$

The reformation of peroxy radicals in Eqs. 8 or 10 constitutes the chain-carrying reaction during which the hydrocarbons in the polluted air are oxidized and NO is converted to NO_2. The following diagram summarizes the events in the catalytic cycle of photochemical smog:

$$
\begin{array}{ccc}
NO + \begin{bmatrix} RCOO_2 \\ RO_2 \\ HO_2 \end{bmatrix} & \xleftarrow{\quad (7,\,8,\,9,\,10) \quad} & \begin{bmatrix} RH \\ RCHO \\ CO \end{bmatrix} \\[2em]
\searrow \scriptstyle(6) & & \nearrow \\[1em]
O_3 + NO \underset{(4)}{\overset{h\nu\,(1)}{\rightleftharpoons}} NO_2 + & \begin{bmatrix} RCO_2 \\ RO \\ OH \end{bmatrix} &
\end{array}
$$

The numbering of the reaction scheme refers to the various preceding reactions. The addition of any of several organic species to the NO_x-ozone system accelerates the production of additional ozone. Organics that enhance the oxidation rate include hydrocarbons, aldehydes and ketones as well as halogenated hydrocarbons. Sources for these anthropogenic emissions are:

- unburned and partially burned gasoline from automobile exhaust,
- filling of gas tanks displacing air saturated with gasoline,
- organic solvents used in industry, dry cleaning and paints.

Compared to the role of the hydrocarbons in the NO to NO_2 conversion via Eqs. 7 and 8 the contribution of CO_2 in the reaction sequences of Eqs. 9 and 10 is likely to be minor. Thus, the rate and efficiency of the production of RO_2, HO_2, and $RCOO_2$ radicals largely determines the ozone level in photochemical smog. Clearly, the rate of reaction of the OH radical with the different hydrocarbons present in the polluted air (Eq. 7) is a key reaction in the catalytic cycle of photochemical smog. The thousands of volatile organic compounds released from the many sources lead to hundreds of different oxidation and degradation products in the photochemical smog process. The potential of a given hydrocarbon to produce additional ozone in the diagram of the photochemical smog cycle depends largely on the rate of the reaction of OH radicals with the different organic compounds. Before we discuss this aspect in detail we will have a closer look at the sources of the chain carrier radicals in Eq. 6.

Sources of OH and HO_2 Radicals

The very reactive hydroxyl radical is produced in several processes. As shown in Fig. 4, ozone absorbs light in the UV from 360 nm to the solar cutoff near 290 nm. A small fraction of the ozone produced in the course of the photochemical oxidation reactions will photolyse in the sunlight to produce excited and ground-state oxygen atoms, $O[^1D]$ and $O[^3P]$, as well as singlet oxygen molecules, $O_2[^1\Delta g]$

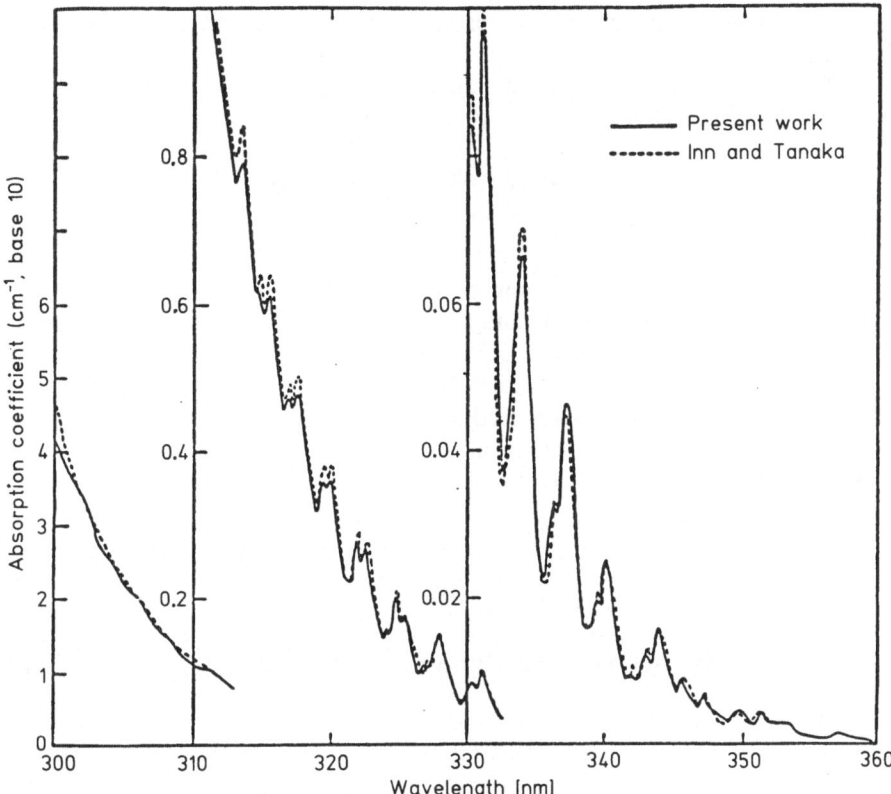

Fig. 4. UV-absorption spectrum of ozone below 360 nm [31]

and ground-state oxygen, $O_2[^3\Sigma g^+]$:

$$O_3 + h\nu(290\text{–}360 \text{ nm}) \rightarrow O[^1D, {}^3P] + O_2[^1\varDelta g, {}^3\Sigma g^+]. \qquad (12)$$

It has been suggested some 15 years ago that singlet oxygen molecules, an electronically excited species of molecular oxygen, play a significant role in the rapid and complex oxidation of NO to NO_2 [32]. The reaction rate of singlet oxygen with common air pollutants and halogenated hydrocarbons [33, 34], however, is too slow to compete with the radicals in Eq. 6 which are responsible for the rapid $NO \rightarrow NO_2$ conversion. In the wavelength range below 319 nm exclusively $O[^1D]$ atoms are produced. The rate coefficient for Eq. 12 can be calculated from the O_3 absorption coefficients (see Fig. 4) as a function of the wavelengths, the quantum yield of the ozone photolysis and the solar actinic flux for a given solar zenith angle in the same manner as was explained for the photolysis rate of NO_2 in Eq. 2. The majority of the very short-lived $O[^1D]$ atoms will be quenched to $O[^3P]$ by collision with nitrogen or oxygen molecules in the atmosphere. However, in collision with a water vapour molecule, $O[^1D]$ reacts readily to form OH radicals:

$$O[^1D] + H_2O \rightarrow 2OH. \qquad (13)$$

Despite the low concentration of water vapour in the atmosphere, at 50% relative humidity 4.5% of the $O[^1D]$ atoms react in a very fast reaction to form OH radicals [35]. Other sources for the production of OH radicals stem from the photolysis of nitrous acid (HONO) and probably hydrogen peroxide (H_2O_2):

$$HONO + hv \ (\lambda < 400 \ nm) \ \rightarrow \ HO + NO \qquad (14)$$

$$H_2O_2 \ + hv \ (\lambda < 370 \ nm) \ \rightarrow \ 2\,HO \qquad (15)$$

which have been built up in the course of the photochemical smog process. Nitrous acid is a good absorber of sunlight for wavelengths lower than 400 nm.

A candidate for the generation of HO_2 radicals in the urban atmosphere is the photolysis of aldehydes, in particular formaldehyde and acetaldehyde. As the photochemical smog process proceeds, these aldehydes are produced from the many oxidation processes of hydrocarbons (see Eq. 11). The absorption spectra of these aldehydes stretch out farther towards the visible region. Also the actinic flux increases greatly at these longer wavelengths (see Fig. 5).

$$HCHO + hv \ (\lambda < 370 \ nm) \ \longrightarrow \ HCO + H \qquad (16)$$
$$\longrightarrow \ H_2 \ + CO$$

$$RCHO + hv \ (\lambda < 350 \ nm) \ \longrightarrow \ R \ + CHO \,. \qquad (17)$$

The photodissociation of formaldehyde into two reaction channels (see Eq. 16), and the complex dependence of the quantum yields on pressure, temperature and wavelength of excitation make it difficult to extrapolate laboratory results to tropospheric conditions. A careful review of the photochemistry of formaldehyde and the extensive work in this field have been summarized and discussed by Calvert [37].

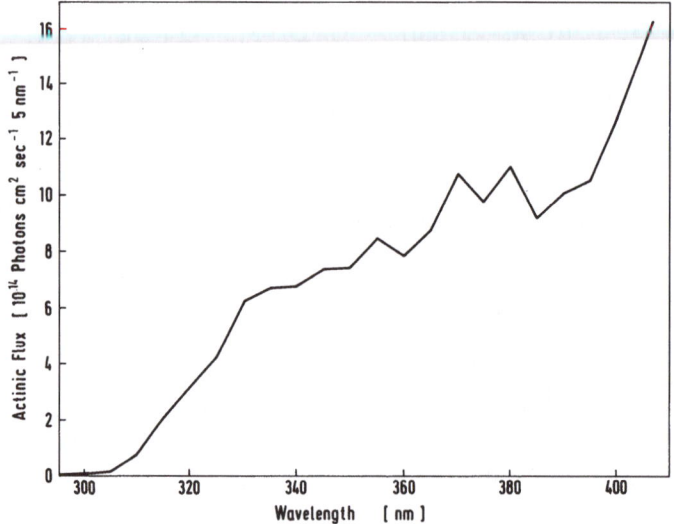

Fig. 5. Actinic solar flux for a solar zenith angle of 40° as a function of the wavelength [36]

Table 1. Photochemical radical sources in the atmosphere, wavelength range and photodissociation rate coefficient J for central Europe at summer solstice [38]

Photochemical process, consecutive Eq. No.	Wavelength range [nm]	$J\ [s^{-1}]$
$NO_2 + h\nu \rightarrow NO + O$ Eq. 1	400–290	0.0078
$HONO + h\nu \rightarrow OH + NO$ Eq. 14	390–290	ca. 0.001
$HCHO + h\nu \rightarrow H + HCO$ Eq. 16	330–290	1.5×10^{-5}
$H_3CCHO + h\nu \rightarrow CH_3 + CHO$ Eq. 17	320–290	3.4×10^{-6}
$O_3 + h\nu \rightarrow O[^1D] + O_2$ Eq. 12	319–290	ca. 2×10^{-5}

In air, the formyl radical HCO, as well as the hydrogen atom produced in Eq. 16 subsequently may form HO_2 radicals:

$$HCO + O_2 \rightarrow HO_2 + CO \tag{18}$$

$$H \quad + O_2 \rightarrow HO_2 \tag{19}$$

In Table 1 the most important radical sources for the photochemical generation of the active OH and HO_2 radicals in the catalytic NO to NO_2 conversion are summarized.

Significance of the Hydrocarbon/NO$_x$ Ratio

As has been outlined in the previous sections, the net formation of oxidants in the photochemical smog requires NO_x, reactive hydrocarbons and sunlight. Under different ambient conditions the rate of ozone formation can be limited by the availability of either NO_x or reactive hydrocarbons in the polluted air. In general, among the reactive hydrocarbons, all hydrocarbons are summarized with the exception of methane. The term NMHC, *non-methane hydrocarbons*, is often used in this context. Thus, the ratio of reactive NMHC to NO_x is of critical importance in the formation and, hence, in the control of ozone. The significance of the NMHC/NO_x ratio in the ozone formation is shown in Fig. 6. This is a diagram of so-called isopleths of maximum afternoon O_3 concentration as a function of the early morning NMHC and NO_x concentrations. Ambient air measurements and smog chamber studies have shown that the peak ozone concentrations occur several hours after the emission of the precursors NO_x and NMHC (see Fig. 1). The isopleth diagram is derived from a chemical kinetics model of photochemical smog which was tested in smog chamber studies by irradiating automobile exhaust mixtures with changing NMHC/NO_x ratios at constant light intensity.

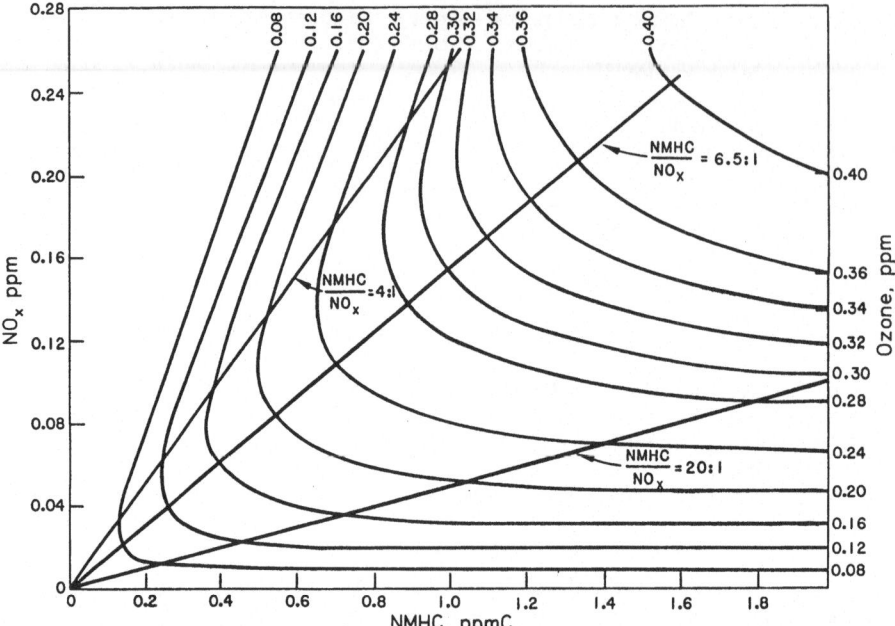

Fig. 6. Isopleth plot of the dependency of maximum afternoon ozone concentration to morning precursor concentrations of NO_x/NMHC

This concept was developed by Dimitriades [39] in a 2.8 m³ smog chamber. Since then the isopleth concept has been used succesfully by other authors to illustrate the results in experiments simulating the photochemical smog chemistry in smog chambers [40, 41]. When the starting concentrations of NO_x and NMHC are plotted on the ordinate and the abscissa, the resulting maximum ozone concentrations in the photochemical smog can be projected as contour lines in the isopleth diagram. After projection to ambient air situations, the following conclusions can be drawn from Fig. 6:

- Reduction of the NMHC emission, i.e. moving horizontally towards the left in Fig. 6, will yield ozone reduction. In most urban areas where the ambient NMHC/NO_x ratio is less than about 10:1, NMHC reductions are most effective in reducing ozone levels. In remote areas where the ambient NMHC/NO_x ratio is about 30:1 or above, increases or decreases in ambient NMHC concentrations may have an insignificant effect on O_3 concentrations.
- Reduction of NO_x, i.e. moving vertically downwards in Fig. 6, can lead to ozone reductions and increases, depending on the NMHC/NO_x ratios. At a low NMHC/NO_x ratio typical of an urban area, allowing NO_x to increase can lower the ozone concentration locally through the NO "titration reaction" (Eq. 4). The downwind transport of the NO_x into a rural area with a typical NMHC/NO_x ratio can increase the ozone concentration. At high NMHC/NO_x ratios, reducing NO_x will be most effective in reducing the ozone level.

The potential effect of NO_x reductions in source areas on the downwind concentrations of ozone and other oxidants is the subject of scientific debate. It is surprising that despite the extreme complexity of the air pollutant mixture in urban air, the general features outlined above can account reasonably well for laboratory and field observations. To illustrate this complexity more than 850 different gas-chromatographic peaks have been resolved in a sample of unleaded gasoline [42]. Gasoline is a complex mixture of pure hydrocarbons with the three most common classes of paraffins, olefines and aromatics with a total concentration of nearly 90%. It is clear that it is impossible to investigate in smog chambers the detailed chemistry of all individual compounds which are discharged into the air. Although smog chamber studies on selected organic solvents such as the chlorinated ethylenes, which are frequently used in the polymer industry and in dry cleaning, have been tested with a view to their behaviour and fate in the photochemical air pollution system [43], more answers to this question can be expected from tropospheric chemical reaction schemes "investigated" on a computer.

Computer Modelling of the Photochemical Smog

As indicated in the preceding section, the chemistry of the polluted air is extremely complex, mainly because of the structural diversity of the non-methane hydrocarbons (NMHC) which occur at variable mixing ratios. Mathematical computer model development based on generalized reaction schemes has progressed through the last 15 years. A photochemical smog model with generalized reaction schemes can be used to dissect the complex series of coupled reactions and learn something about the influence of selected pollutants within this smog chemistry. The quality of the computer model for evaluating the change of primary pollutant concentration and ozone formation with irradiation time depends on the proper representation of the complex reaction scheme, i.e. in the selection of:
- all primary pollutants and intermediate radicals involved in photosmog chemistry,
- all and the most reliable reaction rate constants for the reactions involved in the scheme,
- the proper starting conditions such as initial concentration of the primary pollutants, temperature, intensity and spectral distribution of the incident light, etc.
During the last 15 years two distinct classes of computer models have been developed.

1. Detailed Models

They incorporate as many primary pollutants and organic trace gases as possible, using individual reaction and degradation schemes for each. All relevant reaction rate constants, reasonable estimates of unknown reaction rate constants and branching ratios of radical reactions as well as of the radical termination reactions have to be incorporated in the reaction scheme. The first detailed model was developed by Niki et al. [4] for computer simulation of the smog chamber experiments of the system NO_x/propene. Demerjian et al. [3] extended the detailed model by incorporating more than one primary pollutant. Methods are given for the estimation of reaction rate constants which have not been measured yet. Com-

puter modelling of photochemical air pollution formation in urban areas and in
England have been developed by Hov et al. [44] and Derwent and Hov [45]. These
detailed models allow the calculation of hitherto unknown reaction rate constants
in a reaction scheme of coupled radical reactions. The necessarily very large
number of individual reactions consumes much computer time. Furthermore, due
to the very long time required for computing, these urban photochemical smog
models cannot be integrated in detailed transport models which are used to predict the oxidant concentration after transport from urban to rural areas.

2. Lumped-Parameter Models

They incorporate a fairly complete set of all the inorganic reactions involved in
photosmog chemistry. The organic chemistry, however, is reduced to a few sym-
bolic hydrocarbons and aldehydes. This keeps the number of reactions and reactants reasonably low. The chemical reaction scheme comprises only the possible
reactions of the primary pollutants NO_x and total hydrocarbons with the reactive
intermediates such as the OH, HO_2, RO_2, and RCO_2 radicals and the oxygen
atoms. Thus, in principle, the following differential equations of the material balance have to be solved:

$$\frac{d[NO_2]}{dt} = -k_1[NO_2] + k_4[NO][O_3] + k_6[NO][RO_2] - k[NO_2][OH]$$
$$-k_{26}[NO_2][RCOO_2]$$

$$\frac{d[NO]}{dt} = k_1[NO_2] - k_4[NO][O_3] - k_6[NO][RO_2] - k[NO][OH] + /\overline{NO}$$

$$\frac{d[HC]}{dt} = -k[HC][O] - k[HC][OH] - k[HC][O_3] - k[HC][RO_2] + /\overline{HC}$$

The differential equations summarize all possible reactions which destroy or
build up NO_2, NO and hydrocarbons, respectively. The numbering of the raction
rate constants k refer to the numerical order of the reactions in this work.

$/\overline{NO}$ and $/\overline{HC}$ are production terms for the primary pollutants NO and hydrocarbons from sources. Chemical symbols in square brackets denote concentra-
tions in molecules per cm^3. The concentration of the radicals are steady-state concentrations. The first lumped-parameter models have been developed for the urban area of Los Angeles by Eschenroeder and Martinez [46] and Hecht et al. [47].
The model of the former investigators is based on computing the concentration
changes that occur in a hypothetical vertical column of polluted air. This so-called
trajectory model predicts concentration changes along a chosen wind trajectory.
The input data are restricted to twelve chemical reactions only, essentially the
reactions of Eqs. 1–7, the photolysis of HONO (see Table 1), and the reaction of
$O[^3P]$ atoms and O_3 with hydrocarbons. Table 2 compiles the reaction of the
lumped-parameter mechanism of Eschenroeder and Martinez [46].

A comparison of the twelve reactions in Table 2 with the same reaction in
Table 1 reveals that there is a difference of a factor 60. This is because the reaction
rates in Table 2 have been adjusted to verify the concentration-time profile of
NO, NO_2, and O_3 with data from smog chamber studies. Logically, these lumped
models are not able to predict reaction rate constants of hitherto not measured

Table 2. Chemical reactions of the lumped-parameter model of Eschenroeder and Martinez [46]

Reaction (Eq.)			Eq. No. [a]	Rate constant [b]
1. NO_2	$+ h\nu$	$\rightarrow NO + O$	1	$0.4\,min^{-1}$
2. O	$+ O_2(+M)$	$\rightarrow O_3(+M)$	3	$2.64 \times 10^6\,min^{-1}$
3. O_3	$+ NO$	$\rightarrow NO_2 + O_2$	4	$40\,ppm^{-1}\,min^{-1}$
4. O	$+ HC$	$\rightarrow 2\,RO_2$		$6{,}100\,ppm^{-1}\,min^{-1}$
5. OH	$+ HC$	$\rightarrow 2\,RO_2$	7, 8	$80\,ppm^{-1}\,min^{-1}$
6. RO_2	$+ NO$	$\rightarrow NO_2 + RO$	6	$1{,}500\,ppm^{-1}\,min^{-1}$
7. ROO_2	$+ NO_2$	$\rightarrow PAN$	26	$6\,ppm^{-1}\,min^{-1}$
8. OH	$+ NO$	$\rightarrow HONO$		$10\,ppm^{-1}\,min^{-1}$
9. OH	$+ NO_2$	$\rightarrow HNO_3$		$30\,ppm^{-1}\,min^{-1}$
10. O_3	$+ HC$	$\rightarrow RO_2$		$0.0125\,ppm^{-1}\,min^{-1}$
11. NO	$+ NO_2(+H_2O)$	$\rightarrow 2\,HONO$		$0.01\,ppm^{-1}\,min^{-1}$
12. $HONO$	$+ h\nu$	$\rightarrow OH + NO$	14	$0.001\,min^{-1}$

[a] The number refers to the numerical order of the reactions in this work.

[b] The second order rate coefficients in computer models are usually given in $ppm^{-1}\,min^{-1}$ instead of the physically more meaningful $cm^3 \cdot molecule^{-1} \cdot sec^{-1}$ dimension. The former dimension depends on temperature and pressure. To convert $ppm^{-1}\,min^{-1}$ into $cm^3 \cdot molecule^{-1} \cdot sec^{-1}$ the former has to be multiplied by the factor 1.87×10^{-13} (p, T). p and T are the pressure in bar and T the temperature in degrees Kelvin.

reaction rates in a proposed reaction scheme. A new advanced lumped-parameter model of Falls et al. [48] operates with reaction rate constants measured precisely, and the total hydrocarbons are divided into sub-classes such as alkanes, olefins, aromatics and aldehydes. A new approach of a lumped model is the so-called Carbon-Bond-Mechanism [49], a very comprehensive lumped-parameter model to simulate the formation and distribution of ozone in the Los Angeles metropolitan area. A critical evaluation of the latter model was published by Schurath et al. [50].

These mathematical modelling of the chemical events of photochemical air pollution can be evaluated by comparing their predictions with the concentration-time profile of smog chamber experiments. Since, however, it is known that there exists a long-range transport of the air parcels of a metropolitan area contaminated with photochemical oxidants, the question of the transport and final fate of these reactive compounds is of primary interest. Thus, in addition to the chemical reactions as a function of the concentration of the various pollutants and reactive radicals, a transport model is required taking into account meteorological parameters such as wind speed and direction at each location in the region as a function of time, the vertical temperature profile and the radiation intensity. Special transport models for photochemical air pollution have been developed. In connection with transport models there is a higher demand on the quality of the chemical models. Thus, lumped-parameter photochemical smog kinetic models cannot be used in this context. They will be treated in the section on long-range transport.

As mentioned before, more advanced lumped-parameter models are used in the attempt to differentiate between the different classes of hydrocarbons. In the real atmosphere of urban areas, however, other volatile organic compounds from sources other than automobiles are present. Because of the very large number of individual pollutants, some of which are known as cancerogenic or mutagenic, it is not possible to carry out relevant smog chamber tests or computer modelling with a view to validating their role in the photochemical smog process. To estimate the reactivity of an individual organic compound in the photochemical smog process another approach is necessary.

The Reactivity of the OH Radical

The reaction of the OH radical triggers an avalanche of radical reactions in the photochemical smog cycle. Unlike many other organic radicals the OH radical is unreactive with respect to atmospheric oxygen. It lives long enough in polluted air to react with compounds present as impurities. Its lifetime in air is about one second. Thus, for a quantitative discussion of the tropospheric behaviour of individual compounds the rate coefficient of the OH radical with a given compound allows to assess its reactivity in the photo-oxidation process. The reactivity of this reaction determines whether a given compound will be oxidized in the urban air parcel or if it can escape from it into rural areas. Since the process of OH radical formation (Eqs. 12 and 13) is also operative in unpolluted air there is a steady-state concentration of this highly reactive radical available which finally determines the lifetime of volatile pollutants in the troposphere. Recently, ground-level OH concentrations have been measured in the troposphere by different methods. The measured steady-state concentration range is between 10^6 to 10^7 OH radicals per cm^3. A summary of the measured OH radical concentrations has been given by Hübler et al. [51]. It is interesting to note that in smoggy air obviously lower OH concentrations are present than in less polluted air. Since among the reactions in the homogeneous gas phase the reaction of the OH radical with a compound is the most important degradation path, the tropospheric half-life, $\tau_{1/2}$, can be calculated by assuming a pseudo-first-order behaviour:

$$\tau_{1/2} = \frac{\ln 2}{k_{OH} \cdot [OH]} \quad (20)$$

Thus, besides the knowledge of an average global tropospheric OH radical concentration the knowledge of the OH rate coefficient, k_{OH}, is of primary importance. Much efforts have been spent during the last decade to develop experimental techniques to determine k_{OH}. A recent review by Atkinson et al. [52] compiled about 150 OH reaction rate constants as well as the different measurement techniques. In view of the large number of chemical compounds concerned, theoretical or empirical methods of predicting the hydroxyl reactivity have been developed. Room temperature rate constants for the gas-phase reaction of OH radicals with organic compounds can be estimated by means of a statistically significant correlation with the corresponding rate constants measured in liquid water [53]. The larger body of k_{OH} data available in water [54], however, allows a rapid estimate of k_{OH} for only those data which have been measured. Thus there is a strong desire for purely theoretical or semi-empirical methods deducing k_{OH}. Zetzsch [55] used structure reactivity relationships such as the Hammett equation to predict OH reaction rate constants. Güsten et al. [56] have shown that there

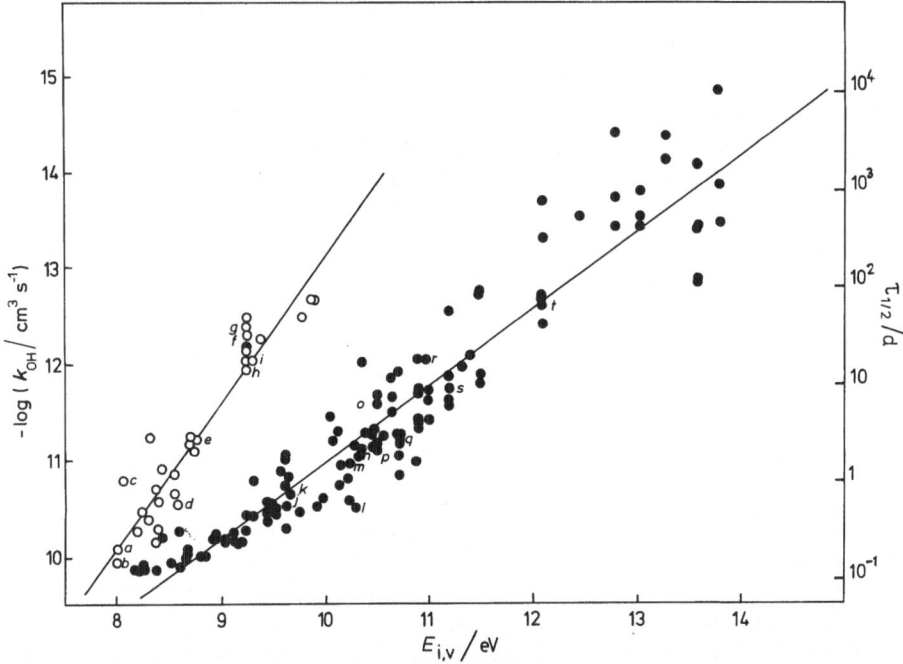

Fig. 7. Correlation of the negative logarithm of the OH reaction rate coefficient, $-\log k_{OH}$, to the vertical ionization energies, $E_{i,v}$, for 161 organic compounds in the gas phase at 300 K [56]

exists a statistically relevant correlation between the reaction rate coefficient, k_{OH} for the OH radical reaction with 161 organic compounds in the gas-phase at 300 K and the corresponding vertical ionization energies. Figure 7 shows the classification of the correlation of $-\log k_{OH}$ to the vertical ionization energy, $E_{i,v}$, for 161 organic compounds in the gas-phase. The correlation reveals two classes of organic compounds only, aromatic and aliphatic compounds. Figure 7 results in the two linear equations:

– $-\log(k_{OH}/cm^3\ s^{-1}) \cong 3/2 E_{i,v}(eV) - 2$ for aromatic compounds and
– $-\log(k_{OH}/cm^3\ s^{-1}) \cong 4/5 E_{i,v}(eV) + 3$ for all aliphatic compounds.

The two linear equations of the OH radical reactivity in Fig. 7 obviously reflect the alternative reaction paths fo the OH radical at 300 K, namely:

– OH addition to aromatic π-systems

$$HO\cdot\ +\ \underset{}{\bigcirc}{-}X \longrightarrow \underset{X}{\bigcirc}\overset{H\ OH}{\cdot} \xrightarrow[\text{air}]{OX\ (-H)} \underset{X}{\bigcirc}{-}OH$$

– H abstraction from aliphatic compounds

$$RH + OH \rightarrow R + H_2O.$$

Since ionization energies are much easier and more rapid to measure than the OH rate coefficients requiring costly and time consuming measurement techniques, the two linear equations of Fig. 7 allow to predict the rate coefficient, k_{OH}, with a probability of 90%.

With Eq. 20 and assuming a global diurnal mean of the OH radical concentration of 5×10^5 cm^3 [57], the upper limit of the tropospheric half-life of organic chemical compounds and, hence, their abiotic persistence can be estimated. This is indicated in Fig. 7 on the right-hand side ordinate, where the tropospheric half-life is plotted. The average OH radical concentration of 5×10^5 cm^3 is a value which has been derived from model calculations for remote, non-urban areas. In polluted urban air, however, higher OH radical concentrations have been measured [51]. It can be seen from Fig. 7 that the lifetimes of organic compounds cover a wide range from a few hours up to years. In particular, the halogenated alkanes (full points on the right side above $\tau = 10^2$ d in Fig. 7) have long atmospheric lifetimes between three months and about thirty years.

General Mechanism of the Oxidation of Hydrocarbons

The general features of the photochemical smog process consist in consecutive chain reations in which atoms and free radicals participate in the many oxidation reactions. In the terminology of chain reaction processes the reactions in photochemical smog can be classified as follows:

1. *Chain initiation*. These reactions comprise all the free radical sources, mainly the primary photochemical processes, which have been described in the preceding sections and summarized in Table 1.

2. *Chain propagation*. In these reactions there is no net gain or loss of radicals. Some of the inorganic radicals produce organic radicals, as shown in the reaction sequence of Eqs. 7, 8. Here, the attack of the OH radical on the various classes of hydrocarbons is the first step in their oxidation. The large oxidation potential of the molecular oxygen of the troposphere is used to form peroxy radicals (see Eqs. 8 and 11).

3. *Chain branching*. In general, there is a net increase in radical species in these reactions, for example:

$$O + \text{hydrocarbon} \rightarrow \text{product} + \text{radicals}$$

with more than one radical produced. In the polluted air a very small fraction of the ground-state oxygen atoms produced in Eq. 1 do not form ozone (Eq. 3) but react with the unburned hydrocarbons, in particular with olefins.

$$O[^3P] + \begin{array}{c} R_1 \\ \diagdown \\ \diagup \\ R_2 \end{array} C = C \begin{array}{c} R_3 \\ \diagdown \\ R_4 \end{array} \rightarrow [\text{adduct}] \rightarrow R_3 - \overset{\overset{\displaystyle R_2}{|}}{\underset{\underset{\displaystyle R_4}{|}}{C}} \cdot + R_1 - \dot{C} = O. \qquad (21)$$

Depending on the chemical structure of R_1—R_4, the fission of the olefin-$O[^3P]$ – adduct may result in other radical formations, e.g. R_1, R_2, R_3—C· and $R_4C = 0$. When the oxidation process of the unburned hydrocarbons has proceeded, the oxygen atoms might react with their oxidation products, e.g. the aldehydes formed in the oxidation processes:

$$O[^3P] + RCHO \rightarrow R - \dot{C} = 0 + OH. \qquad (22)$$

In any event, the result of the branching reactions is the formation of two free radicals.

4. *Chain termination.* In these reactions free radicals are removed from the highly reactive mixture through the formation of stable end products such as aldehydes, ketones, PAN, nitric and nitrous acids, etc.

Due to the chain nature of all these oxidation reactions, the free radical concentrations are maintained in a steady state depending on their production rate in the various primary photochemical processes (see Table 1), their conversion rates in the chain propagation and chain branching reactions, and their removal from the polluted atmosphere in chain termination reactions.

Products of the Photochemical Smog

When discussing the various products formed in the photochemical smog cycle we have to bear in mind that these products are generally more stable than the many other compounds in the chain reactions. This allows to detect them analytically and determine them quantitatively. A second point is that nearly all the smog products can be found in rural and clean remote areas, i.e. they are also of natural origin. Up to now we have no means of differentiating between natural or anthropogenic air pollution products. Only measurement of their actual concentrations in urban air and comparison with their natural "background" concentrations gives a reliable estimate of the anthropogenic pollution.

Ozone

In urban areas ground-level ozone is frequently used as an indicator of photochemical smog. According to Fig. 3, showing the urban diurnal cycle of NO_x and ozone, ozone reaches a maximum in the early afternoon. Due to Eqs. 1 and 3 this high ozone concentration is indoubtedly "anthropogenic" ozone. According to Singh et al. [58] typical ozone concentrations in clean remote atmospheres range from 20–40 ppb. In general, it is believed that there are two natural sources of ozone. In spite of the very slow and steady gas exchange between the stratosphere and the troposphere, irregular intrusions of stratopsheric ozone through events called "tropopause foldings" seem to be the main source of global tropospheric ozone. Aircraft measurements of ozone at 8 km height indicated that intrusion of ozone occurs predominantly in tropospheric regions of low pressure [59]. A second source of ozone in the clean remote troposphere is believed to result from the reaction sequence Eq. 9, 6, 1, 3, which is in equilibrium with the ozone-destroying photochemical reaction sequence Eq. 12, 13, followed by:

$$OH + O_3 \rightarrow HO_2 + O_2 \tag{23}$$

and

$$HO_2 + O_3 \rightarrow 2O_2 + OH. \tag{24}$$

On the basis of vertical concentration profiles of NO_x and CO measured in the troposphere [60, 61] – the stable atmospheric gases in the ozone buildup reaction sequence Eq. 9, 6, 1, 3 – global model calculations indicate a positive ozone concentration for the Northern Hemisphere between the ozone-destroying reaction sequence Eq. 12, 13, 23, 24 and the ozone-forming reaction sequence Eq. 9, 6, 1, 3 [62–65]. Of course, this in-situ photochemical source of global tropospheric ozone needs further experimental confirmation. This ozone source is difficult to sort out from the intrusion of stratospheric ozone. At any rate, the theoretical model consideration indicates an average global "background" ozone concentration of 20–40 ppb, which corresponds quite well to the concentrations measured in the clean remote troposphere [58].

The measurement of ozone in meteorological stations and remote areas has a long tradition in Germany [66–68]. However, in urban areas, ozone was measured in Frankfurt as late as in 1967 [69] and in Berlin in 1969 [70]. While the data measured in Berlin were interpreted in the sense that they did not favour the formation of photochemical oxidants, Jost [70] discussed the formation of higher ozone levels during the afternoon as generated by the photochemical smog cycle. This led to discussions whether photochemical smog could be generated in Europe [71]. It was concluded that due to less insolation, photochemical smog formation was a minor problem in Europa north of the Alps [70, 71]. However, elevated ozone levels in the Netherlands [72], southern England [73] and London [74] measured during anticyclonic periods in the summer, left little doubt that ozone with values exceeding 120 ppb over the daytime makes a significant contribution to photochemical smog. During 1974/75 a large programme was started to measure ozone and its precursors, mainly NO_x and hydrocarbons, in Germany, in particular in Bonn [75], Cologne [76], Frankfurt [77] and Karlsruhe [78]. Since then, numerous measuring sites for atmospheric pollutants including ozone have been built up by meteorological, hygienic and control institutions. In Table 3 the annual maximum ozone values and the average ozone concentrations for the period of April to September – the period of high insolation – have been compiled from a selected number of stations in the Federal Republic of Germany. More detailed lists of ozone concentration data from nearly all stations and measuring sites in Germany can be found in Ref. [38]. The ozone concentration data in Table 3 provide an overview of the widespread surface ozone concentrations in Germany at different altitudes. Table 3 reveals the following:

- While the maximum ozone concentrations in cities are high, the average ozone concentration over the summer is low compared to measuring sites located at higher altitudes. This is particulary evident from the comparison of the data from Bonn with the data from the Ölberg nearby. The reason ist that during the night, ozone is depleted in the cities due to anthropogenic emissions of NO_x and hydrocarbons.
- At very high altitudes, as on the Hohenpeißenberg (~1,000 m) and the highest mountain in Germany, the Zugspitze (2,964 m), the maximum ozone concentrations and the average ozone concentrations over the summers are nearly constant throughout the years.
- Very high maximum ozone concentrations and relatively high average ozone concentrations were recorded in the years 1976 and 1983 during the two very hot summers with unusual meteorological conditions favouring the photochemical production of ozone.
- The highest ozone concentrations are generally measured in the Rhine Valley in the cities of Mannheim and Karlsruhe.

Table 3. Ozone concentrations ($\mu g/m^3$) * in the Federal Republic of Germany from 1975–1983
$O_{3(max)}$: annual maximum ozone concentration
ΣO_3: average ozone concentration for the period of April to September

Station	1975 $O_{3(max)}$	ΣO_3	1976 $O_{3(max)}$	ΣO_3	1977 $O_{3(max)}$	ΣO_3	1978 $O_{3(max)}$	ΣO_3	1979 $O_{3(max)}$	ΣO_3	1980 $O_{3(max)}$	ΣO_3	1981 $O_{3(max)}$	ΣO_3	1982 $O_{3(max)}$	ΣO_3	1983 $O_{3(max)}$	ΣO_3
Cologne[a]	280	36	390	38	300	28	250	66	190	32	240	36	546	50	326	52	304	50
Bonn[a]	320	46	370	42	396	30	348	28	221	–	274	–	–	–	–	–	–	–
Ölberg[a] (460 m)	260	62	340	78	260	56	340	68	350	66	250	60	318	64	–	–	–	–
Mannheim[b]	298	57	543	53	255	43	237	34	244	46	193	38	257	44	408	57	187	53
Karlsruhe[b]	–	–	358	46	409	–	299	44	200	39	287	33	245	35	260	59	222	39
Frankfurt[a]	–	–	409	68	211	32	231	34	278	35	315	48	347	43	366	54	270	40
Hohenpeißenberg[c] (~1,000 m)	202	83	184	86	170	74	198	80	188	74	166	73	168	78	240	86	182	82
Zugspitze[c] (2,964 m)	–	–	–	–	113	60	87	54	106	60	110	61	121	65	137	77	181	106

* $2 \mu g/m^3 \cong 1$ ppb ozone
a Half hour average ozone concentration
b Three hour average ozone concentration
c One hour average ozone concentration

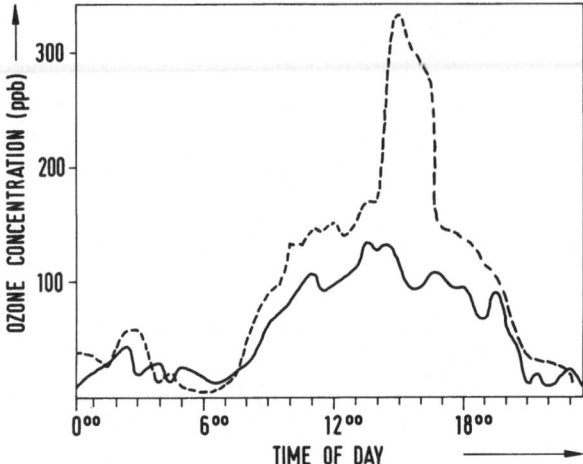

Fig. 8. Diurnal variation of the ozone concentration in Mannheim (– – –) and Karlsruhe (———)
on June 23, 1976 [79]

One of the highest mean hourly average volume fractions of ozone in the Federal Republic of Germany was measured in Mannheim during the exceptionally hot summer 1976 [79] (see Fig. 8). Synoptic conditions over Europa during that summer were dominated by a large high-pressure area which produced sunny skies, high temperatures, and low to moderate wind speeds over central and northwestern Europe. High ozone concentrations are generally also measured in the area of Cologne/Bonn. On August 15, 1981, ozone concentrations of 520 µg/ m^3 were recorded in the industrial area of Cologne for several hours. During the daylight hours, while photochemical smog production takes place, the mixing height increases at the same time, which causes ozone-rich air from aloft to be transported to the ground. For these reasons the maximum ozone concentration is reached in the early afternoon and the diurnal pattern of the ozone concentration is generally independent of the geographic location (see Fig. 1, 3, and 8). The highest ozone concentrations are measured in June and July. All these ozone data have to be compared with the World Health Organisation (WHO) long-term goal for maximum ozone concentrations per hour of 60 ppb. Though ambient air quality standards for most countries are higher, there is no doubt that during the last ten years in many German cities even higher mean ozone concentrations per hour than 120 ppb have been measured frequently. A discussion of the different national air quality standards for ozone is given in a later section of this chapter.

The general question is whether ozone concentrations measured at ground level in cities and industrialized areas are a representative and quantitative indicator of the severity of photochemical smog. In areas near the ground, destruction processes can occur by a catalytic destruction of ozone in contact with the surface and by chemical quenching through reactions in the lowest air layers, mainly by NO and olefinic hydrocarbons. Thus, measuring sites at ground level do not give a representative picture of the widespread horizontal and vertical distributions of

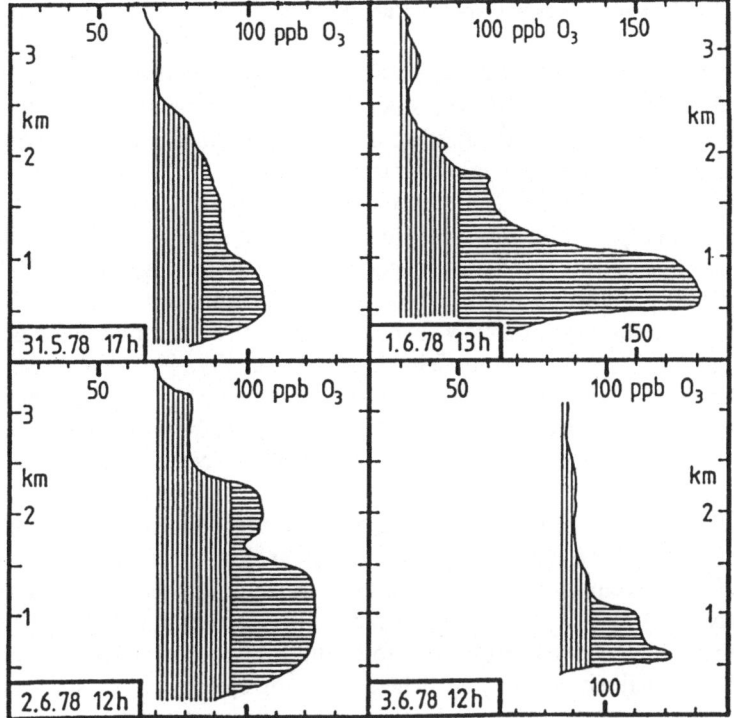

Fig. 9. Vertical ozone concentration profiles in the Cologne-Bonn area on May 31 to June 3, 1978 [82]

▥ supra-regional excess of ozone in the mixing layer
▤ locally produced ozone

ozone. This has been proved by aircraft measurements of horizontal and vertical profiles of the ozone concentrations in Los Angeles [80], California [81], in the Rhine-Ruhr area [82, 83] and in the Netherlands [84]. In Fig. 9 the vertical profiles of the ozone concentrations in the Cologne-Bonn area on four consecutive days indicate the strong influence of micrometeorology on the ozone concentration. Figure 9 clearly indicates that within the mixing layer (planetary boundary layer) higher ozone concentrations can be monitored at a height of several hundred meters. Through convective mixing caused by the warmer earth surface, ozone and its precursors (NO_x and hydrocarbons) are generated and trapped more effectively at several hundred meters height. The ozone concentration within the mixing layer is substantially higher compared to the "background" concentration in the free troposphere. The height of the local inversion layer which lies as a "lid" over the mixing layer, depends on the micrometeorology, i.e. the time-height dependence of wind vectors as well as of temperature. Thus, the ozone concentration monitored at ground-level depends very much on transport phenomena to the ground within the mixing layer. This influence of local meteorology on measured ground-level ozone concentrations have been shown to exist by Muschalik [85].

Ambient Air Concentrations of Ozone in Europe

As indicated in the "Historical background" section, the phenomenon of photo-chemical smog has been observed since 1970 in many large metropolitan areas throughout the world. After first sporadic measurements of "anthropogenic" ozone in Vlaardingen (The Netherlands) in 1968 [86], continuous ozone monitoring has been performed in Delft since 1969. The results were compiled by Guicherit et al. and presented in several reports and publications [72, 86, 87]. Ozone concentrations of 120 ppb were exceeded frequently throughout the 1970s with a maximum hourly value of 270 ppb of ozone in the summer of 1976. In England, regular measurements of ozone were conducted with a chemiluminescence detector in Harwell in 1971 [73] and Central London in 1972 [74]. The highest ozone concentration of 258 ppb was recorded in the summer of 1976 [88]. Even in Scandinavia at more northern latitudes, around 60°N, surprisingly high concentrations of ozone exceeding values of 120 ppb have been measured since 1975. Schjoldager et al. [89, 90] summarized the results in several reports [91] and publications. At least in the summer during the very long solar days at high latitudes, the integrated solar quantum flux over the solar day is quite similar at summer solstice for different latitudes. For example, at summer solstice the length of the solar day in Oslo is about 18.5 h, while in Los Angeles it is about 14 h [92]. Episodically high concentrations of ozone and PAN occur in south Norway and Sweden. The majority of the photochemical smog episodes are associated with a high-pressure area over northern Europe. When the local weather conditions lead to photochemical oxidation reactions in the atmosphere and polluted air masses are transported to Scandinavia from the major precursor source areas in Europe and Great Britain, the most pronounced oxidant episodes occur [93]. This feature illustrates the importance of the long-range transport of atmospheric pollutants.

In France, ground-level ozone concentrations have been monitored since 1976 in Vert-le-Petit, 35 km south of Paris [94]. Though values of up to 215 ppb have been recorded, it was concluded that the ozone was transported to the measuring sites rather than due to local photochemical smog reactions. Thus, without simultaneous recording of the precursor pollutants NO_x and hydrocarbons as well as the wind speed and direction, it is difficult to discriminate between locally generated or transported "anthropogenic" ozone. Lopez et al. [95] reported on ozone concentrations measured in the southwest of France during 1978/79 [95]. Since the summer of 1976 continuous monitoring has been performed in Illmitz, 65 km southeast of Vienna. Particularly high ozone concentrations were recorded between April and September 1979. During that period maximum ozone concentrations per hour higher than 150 ppb were recorded on 90 days; the highest value was 249 ppb on August 15, 1979 [91]. In Table 4 the hourly ozone concentrations over 200 ppb are compiled for the years 1976–1979 and the following countries: Austria, Federal Republic of Germany, Netherlands and United Kingdom.

The ozone data for different central European countries classified by the number of days with hourly ozone concentrations higher than 100 and 150 ppb for the summer months (May–August) in the years 1976–1979 indicate better the occasional occurrence of photochemical smog in central Europe (see Table 5).

Table 4. Hourly ozone concentrations higher than 200 ppb in central Europe during the years 1976–1979 [91]

Station, country	Date	Maximum concentration (ppb)	No. of hours with concentration >200 ppb
Illmitz, Austria	Apr. 15, 1979	205	3
	June 07, 1979	220	2
	June 11, 1979	213	2
	Aug. 15, 1979	249	3
	Aug. 22, 1979	203	2
	Sept. 14, 1979	208	2
Venusberg, Bonn, Federal Republic of Germany	July 12, 1977	202	1
Delft, Netherlands	1976	200	
Vlaardingen, Netherlands	May 08, 1976	270	
Vlissingen, Netherlands	1976	220	
Haamstede, Netherlands	1976	200	
WSL, United Kingdom	July 03, 1976	207	2
Harwell, United Kingdom	July 02, 1976	220	6
	July 03, 1976	220	7
	July 04, 1976	230	4
	July 05, 1976	258	6
	July 06, 1976	204	1
	July 07, 1976	212	2
MRC City, United Kingdom	June 25, 1976	201	2
	June 26, 1976	203	2
	June 27, 1976	200	1
	July 03, 1976	200	1
GLC Teddington, United Kingdom	June 28, 1976	211	1

Table 5. Number of days during which hourly ozone concentrations of 100 and 150 ppb were exceeded in central Europe during the summer months (May–August) from 1976 to 1979 [91]

Country	Reference value (ppb)	1976	1977	1978	1979
Austria	100	5	15	30	115
	150	0	2	1	65
Belgium	100				6
	150				1
Federal Republic of Germany	100	31	6	9	1
	150	6	2	1	0
Finland	100				1
	150				0
Netherlands	100	29	2	3	7
	150	5	0	0	0
Norway	100	6	1	3	4
	150	0	0	0	2
Sweden	100	12	2	14	7
	150	0	0	0	2
United Kingdom	100	28	13	7	3
	150	14	1	0	1

Again, the high concentrations attained during the warm and dry summer of 1976 are evident. Surprising are the very large numbers of high ozone concentrations in Austria in 1979. In Switzerland, regular ozone measurements were started at three stations, Dübendorf, Basel and Sion, in 1981. The data are reported together with concentration data of other atmospheric pollutants by the Swiss Bundesamt für Umweltschutz in Bern.

Scarce ozone data are available from Italy [96] and Spain [97]. In the latter country ozone was measured by the gas chemiluminescence method in Madrid throughout the year 1976 [97]. Compared to the very high ozone concentrations measured during 1976 in Central Europe (see Tables 3 and 4), the ozone concentrations at the station near Madrid are very low throughout the year 1976.

More data of ozone measurements are available from the Balkan Peninsula. In 1975, continuous ozone monitoring with an ethylene-chemiluminescence detector was performed in the city of Zagreb in ambient air at 65 m height within the period from 1st May until 30th September [98]. Daily maximum concentrations of up to 140 ppb were measured on sunny days. The old 1971 U.S. Air Quality Standard admitting 80 ppb ozone as the mean hourly concentration was exceeded on 41 days during the period of measurement. A statistical analysis of the 3310 hourly average ozone volume fractions was correlated with meteorological factors (rain, wind speed and direction, temperature, hourly average insolation) [99]. More than 70% of the values exceeding the old U.S. Air Quality Standard of 80 ppb are correlated with the wind from the industrial area in the southeast part of Zagreb. Yugoslavian ozone concentration data are further available from the isle of Krk [100] and from the city of Split at the Adriatic coast [101, 102].

Under a German-Greek cooperation programme a measuring campaign was performed in June 1982 in Greater Athens for various pollutants with a mobile laboratory equipped with automatic monitors [103]. On the ca. 250 m high hill of Lycabetos in the centre of Athens, concentrations of up to 300 ppb were measured on June 26. On Lycabetos, after a late evening minimum, the ozone concentration went up again to 150 ppb at midnight of June 25/26. On those days, there was a very strong nighttime inversion as evidenced by a temperature of 27.7 °C at 82 m and 31.2 °C at 540 m above MSL. Thus, large concentrations of ozone generated on the previous day were trapped during the night at some height above the ground, while the strong elevated inversion prevented NO generated by nighttime traffic to reach higher up and destroy it. The phenomenon of nocturnal ozone maxima is most pronounced on mountain tops [104] and in costal regions such as in Jerusalem and Tel-Aviv [105]. During June 1982, very hot weather prevailed for a number of days with temperatures rising to more than 40 °C. Furthermore, the geographic location of Greater Athens (~ 38 °N) in a basin surrounded by high mountains on three sides and the sea on the remaining side gives rise to severe photochemical smog formation. During the summer, on days with anticyclonic weather conditions, symptoms of photochemical smog are the frequently observed strong reduction of the visibility and the appearance of a "brown" cloud over Athens.

Summarizing the overview of the high ozone concentrations in Europe, one reaches the conclusion that high ozone concentrations in ambient air occur in high-pressure situations with anticyclonic weather during the summer.

As the attempt is made in this article to concentrate on the situation of photochemical air pollution in Europe it is beyond the scope and length to report here about the photochemical smog situation on other continents in detail.

The largest body of information concerning ambient concentrations of ozone exists in the USA which has been extensively reviewed [12, 217, 245]. Extensive measurements of ambient ozone concentrations in the St. Louis area during 1974–1977 have been reported [246]. Since it became clear that in many cities and metropolitan areas local generation of ozone and inflow of elevated ozone concentrations due to transport phenomena occur, in the 1970s ozone concentrations were often monitored outside of urban areas, e.g. in rural California [247], St. Louis area [192], and in the northeastern [198] and upper-midwest United States [199].

The occurrence of photochemical smog with high levels of ozone has also been recognized in Canada [248–250], Mexico [251], Japan [14, 252] and Australia [15, 253–255]. Concentrations of up to 200 ppb ozone during sunny anticyclonic weather have been measured in Canadian cities located north of 50° N latitude. In Mexico City, known as one of the most polluted metropolitan areas of the world, a maximum ozone concentration of 470 ppb has been reported [251].

Peroxyacetyl Nitrate (PAN)

Peroxyacetyl nitrate is the principal member of a family of nitrogenous compounds produced in polluted urban atmospheres by photochemical reactions in the photosmog cycle. Acetyl radicals are produced by a variety of reactions in ambient air, including the attack of OH radicals on acetaldehyde. Though the mechanism of formation of PAN and other members of the peroxyacetyl nitrate family has not been studied as extensively as that of ozone, straightforward "paper chemistry" persumes the formation of peroxyacetyl radicals by reactions of ozone with olefins or OH radicals with aldehydes [13, 106]. After rapid formation of the peroxyacetyl radical, PAN is formed in a chain termination reaction with nitrogen dioxide:

$$CH_3CO \quad + O_2 \quad \rightarrow \quad CH_3COO_2, \tag{25}$$

$$CH_3COO_2 + NO_2 \quad \rightleftarrows \quad CH_3COO_2NO_2. \tag{26}$$

The formation of PAN is reversible since the molecule decomposes to give again its precursor species [107]. Thus, PAN can be a temporary sink for NO_2 in sustaining the photochemical smog process. In the absence of sunlight and reactive compounds, PAN can remain in equilibrium for some time. At relatively low temperatures during the night and when air aloft is decoupled from NO surface emissions by an inversion layer, PAN can be transported over large distances and decompose at a downwind location during daylight hours. Decomposition of PAN in remote atmospheres can lead to ozone formation and other manifestations of photochemical air pollution. However, in the presence of NO the equilibrium in Eq. 26 is destroyed by the acetylperoxy radicals reacting preferentially with NO to produce NO_2 (see Eq. 6). Smog chamber studies on the photochemical oxidation of unburned hydrocarbons show that PAN starts to form when the NO concentration is at a minimum (see Fig. 1 b).

As Eq. 26 strongly depends on the temperature, PAN could be very stable in the upper troposphere and lower stratosphere. Recently, Singh and Hanst [108] and Aikin et al. [109] have postulated on the basis of one-dimensional photochemical model calculations that significant PAN concentrations can be synthesized in the colder regions of the unpolluted troposphere. The presence of PAN in the higher troposphere has meanwhile been confirmed by measurements in high altitudes above the Pacific [110]. There is now considerable evidence that PAN may be ubiquitous in the continental boundary layer [114], i.e. present during all seasons and at rural as well as urban locations. Its concentration in rural areas is generally in the sub-ppb-range [114, 115]. Nevertheless, PAN is generally considered to be an urban pollutant.

In urban air PAN is a strong oxidant which is phytotoxic at higher concentrations [111, 112] and can affect human health. PAN is a strong eye irritant [13] and has been suggested as a possible etiological agent in the growing incidences of skin cancer [113]. As PAN strongly reacts with sulfhydryl groups of enzymes and with other sulfur-containing compounds, such as certain amino acids, it is believed that these reactions explain the eye-irritating properties of PAN as well as its toxic effect on plants. Recent laboratory studies have demonstrated that PAN is soluble in acidic water samples and in rain water [116]. PAN decays to nitrate and organic products in the water sample with an aqueous lifetime of a few minutes to more than one hour, depending on ambient temperature. Thus, wet removal could be a significant mechanism of PAN loss in the atmosphere and PAN can be a contributor to nitrate formation during precipitation.

Unlike in rural areas, the PAN concentration in urban air is generally in the low ppb range. A summary of ambient PAN concentrations in North America can be found in Refs. [117] and [118]. The PAN concentration in southern California is five to ten times higher than in North Amerika or Japan. In Riverside,

Table 6. Summary of measurements of ambient PAN concentrations (in ppb) in Europe

City and country	Sampling period	Daily mean*	Highest daily mean	Maximum concentration	Reference
Harwell, England	Sept. 07, 1973	–	–	2.3	122
Delft, Netherlands	June 26–Nov. 02, 1972	0.6	1.4	16	121, 123
Delft, Netherlands	Febr.–Oct. 1975	2	–	20	124
Harwell, England	Nov. 1974–Oct. 1975	1	–	8.9	125
London, England	Aug. 1974–Oct. 1975	–	–	16.1	124
Bonn, F.R. Germany	June 20–Aug. 01, 1975	–	–	1.65	126
Essen, F.R.Germany	June–Sept. 1978	1	–	3.6	127
Riso, Denmark	June 11–Sept. 23, 1980	–	0.9	4.2	128
Goteborg, Sweden	June 17–Sept. 09, 1980	–	0.8	3.5	128
Oslo, Norway	June 04–Sept. 20, 1980 May 21–Aug. 23, 1981	–	–	5.6	129
Klyve, Norway	June 16–Sept. 27, 1982	–	–	13.6	93, 129
Athens, Greece	Sept. 14–Sept. 16, 1982	–	–	0.48	130

* Based on 24 h sampling period

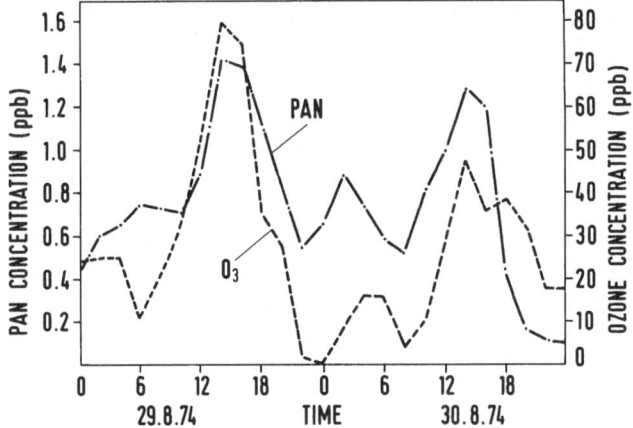

Fig. 10. Time variation of ozone and PAN concentrations at Harwell, England in August 1974 [125]

California, maximum daily concentrations are 20–30 ppb [117]. Contrary to ozone, PAN cannot be measured continuously over the day. Generally, PAN has to be sampled over longer sampling periods in the field and subsequently analysed in the laboratory by electron capture gas chromatography [110]. Laboratory [119] and field compatible calibration procedures for PAN have been described [120]. In Europe, the first measurements of PAN were reported by Nieboer and van Ham [121]. The PAN measurements of the year 1972 were not published before 1976. A linear correlation between the ozone and the PAN concentrations indicates that both secondary pollutants are generated by photochemical smog reactions (see Fig. 10) The average ozone/PAN ratio measured during the daytime is 45, a ratio similar to that measured in the Los Angeles metropolitan area. The seasonal variations of PAN in rural, southeast England and the influence of winter on its concentration has been summarized recently by Brice et al. [131]. A summary of European measurements of ambient PAN concentrations is given in Table 6.

The data in Table 6 indicate that average summer concentrations of PAN in European countries are in the low ppb range and the 10 ppb level is rarely exceeded. Figure 10 shows the typical features of ozone and PAN in ambient air during a period of photochemical smog with the daily maximum occurring in the early afternoon and a nocturnal minimum.

Hydrogen Peroxide (H_2O_2)

Gay and Bufalini observed that H_2O_2 concentrations generally paralleled those of total oxidants during heavy photochemical smog episodes in Los Angeles [132] with concentrations of the order of 180 ppb [3, 133]. In 1978, Kok et al. [134] developed a sensitive method for the determination of hydrogen peroxide in the ambient atmosphere. During days of intensive photochemical air pollution, ozone concentrations of 150 to 200 ppb are monitored; on the average, H_2O_2 concentrations of 10 to 30 ppb were measured in the south coast air basin of California [135]. In the meantime, Zika and Salzman [136] have reported that the analytical method for H_2O_2 determination by Kok et al. [134] is sensitive to ozone. When ozone is present in the air being sampled, there is an anomalous production of H_2O_2 in the traps which adds to the hydrogen peroxide collected from the air. Nevertheless, hydrogen peroxide seems to be a product of photochemical air pollution. However, there seems to be no distinct interdependence between the maximum H_2O_2 concentrations and the maximum ozone concentrations as can be monitored in the case of ozone and PAN (see Fig. 10). It has been suggested that in the troposphere, an H_2O_2 synthesis can occur in the aqueous phase of droplets [137]. This fact might be of some significance since in water H_2O_2 is capable of oxidizing sulfur dioxide to sulfuric acid in acidic solution [138, 139]. As at low pH the SO_2 oxidation by ozone and oxygen is retarded, the SO_2 oxidation by H_2O_2 could be the major process by which atmospheric SO_2 is converted to sulfate (Acid rain problem). There appears to be a growing acceptance to the effect that H_2O_2 results in the termination reaction of two perhydroxyl (HO_2) radicals in photochemical chain reactions of the smog cycle [64, 133, 135):

$$HO_2 + HO_2 \rightarrow H_2O_2 + O_2. \tag{27}$$

Recently, it has been suggested that HO_2 radicals, and hence H_2O_2, can be generated in the liquid phase around an inorganic core of a tropospheric aerosol via heterogeneous photocatalysis [140]. Currently, no published data are available of H_2O_2 concentrations in European cities.

Nitrate Radical (NO_3)

In smog chamber studies of NO_x-O_3-air mixtures [3, 4, 141] the gaseous nitrate radical has long been recognized as a fairly long-living intermediate and recently identified and measured quantitatively by long-path spectroscopic techniques in ambient nighttime atmosphere at various locations in the United States and in Europe [142, 143]. The concentrations measured, range from the detection limit of the spectroscopic technique [~ 1 part per trillion (ppt)] up to about 350 ppt at a downwind receptor site in the Los Angeles Basin [141]. Since the absorption spectrum of NO_3 reaches out over the entire visible part, NO_3 is rapidly photolysed by sunlight to NO_2 and $O[^3P]$ atoms for wavelengths of less than 580 nm. The lifetime of NO_3 at daylight is about 10 seconds only. Furthermore, NO_3 reacts with NO in polluted air. Therefore, NO_3 is most likely detected after sunset when most of the NO has been converted to NO_2. The major process of NO_3 formation is the reaction of NO_2 with ozone produced during the daytime. Indeed, Platt et al. [142, 143] measured maximum NO_3 concentrations within one to two hours after sunset. Significant NO_3 concentrations were also measured at Jülich and Deuselbach, Federal Republic of Germany, in 1980. During a measurement campaign at Deuselbach in April 1980, the NO_3 concentration early in the morning reached 40 to 280 ppt [143].

The reactions associated with the oxidation of atmospheric NO_2 are summarized in the following scheme:

$$NO_2 + O_3 \longrightarrow NO_3 \underset{OH}{\overset{NO_2(M)}{\rightleftharpoons}} N_2O_5 \xrightarrow[(M)]{H_2O} 2\,HNO_3$$

$$OH \atop (-H_2O)$$

$$2\,NO_2 \xleftarrow{\ NO\ }$$

$$NO_2 + O[^3P] \xleftarrow{\ h\nu\ }$$

$$< 580 \text{ nm}$$

(28)

Dinitrogen pentoxide (N_2O_5) can be formed through the reaction of NO_3 with NO_2 in excess. According to the known rate constant for this reaction one can calculate that if the NO_2 concentration is greater than 1 ppb, the lifetime of the NO_3 radical with respect to the reaction with NO_2 is less than one minute. In the presence of water vapour, HNO_3 is formed. The reaction of N_2O_5 with water is supposed to be very rapid. Above a threshold of about 50–60% relative humidity, the lifetime of NO_3 is very short. Under humid conditions NO_3 was generally present below the limits of detection during the night [143]. Even under conditions of dryness the lifetime was only about half an hour. Since in the laboratory the homogeneous gas-phase reaction of N_2O_5 with water vapour is extremely slow [144], it has been speculated that a heterogeneous reaction of N_2O_5 and water on the surface of aerosols readily forming HNO_3 could provide a large sink for NO_2 [145, 146]. However, HNO_3 does not increase in concentration during the night [147] so that there is a lack of direct evidence as to the importance of the N_2O_5 hydrolysis in the real atmosphere. During the day, photochemically generated OH radicals can react with HNO_3 to generate NO_3 which is rapidly photolysed to NO_2 and $O[^3P]$ (see reaction scheme of Eq. 28). Thus, the knowledge of the chemistry of the nitrogen oxides in the lower atmosphere is still incomplete. Recent measurements of rate constants for the gas-phase reactions of NO_3 with organic compounds, together with the ambient atmospheric concentration observed of the NO_3 radical, indicate that the reaction of NO_3 with the more reactive alkenes and olefines of biogenic origin, such as the terpenes, may be an important nighttime sink for both NO_3 radicals and these organic compounds [148, 149]. While the reaction of NO_3 with alkanes is fairly slow [150], the reaction with biogenic organic compounds is fast [151]. In moderately polluted air, terpenes are consumed within a few minutes. Thus, this reaction is equal in importance to loss due to the attack by OH radicals during daylight hours. Compared to the reaction rates of ozone with different organic compounds, NO_3 reacts more rapidly with these organic compounds. The O_3:NO_3 ratio of the reaction with alkanes, alkenes, aromatic coumpounds and aldehydes is $1:2.5 \times 10^2$ to 3×10^4. Though little is known about the reaction products formed in the reaction of NO_3 with organic compounds, it is supposed that the reaction proceeds via the addition of NO_3 to double bonds [148]. Thus, NO_3 could be a source of organic radicals during the night comparable to the reaction of ozone with olefines.

Nitric Acid (HNO₃)

On account of its thermodynamic stability, nitric acid is an important sink for nitrogen oxides in the atmosphere [152]. HNO_3 is a substantial component in acid precipitation, and is apparently rising in proportion to sulfuric acid. In daylight, the reaction of NO_2 with OH radicals is rapid and considered to be the "classical" pathway of HNO_3 formation (Table 2, Eq. 9). Oxidation rates of NO_x on the order of 10 percent per hour have been observed in urban plumes during daylight hours. At night, the heterogeneous source in the $N_2O_5 + H_2O$ reaction (see Scheme 28) is possibly the major non-photochemical source. The rapid reaction of NO_3 via N_2O_5 to HNO_3 (see Scheme 28) possibly occurs on the surface of deliquescent aerosols or moist ground surfaces [145, 146]. As HNO_3 is volatile and does not absorb solar light in the lower troposphere, it is a stable compound and omnipresent in the atmosphere. Therefore, its concentration shows only slight daily variations. Gaseous HNO_3 is obviously rapidly removed from the atmosphere by dry and wet removal processes [153]. As the nitric acid vapour can react with ammonia to form ammonium nitrate, it does not reach high levels in photochemical smog [154]. Ammonium nitrate condenses to an aerosol possibly in association with sulfur-containing aerosols and can lead to high levels of particulate nitrate in the lower troposphere. Thus, average HNO_3 concentrations in urban air are usually below 10 ppb [155]. A summary of ambient air measurements of nitric acid concentrations at urban and suburban rural sites in the United States was prepared by Altshuller [118]. In clean air the concentration of HNO_3 ranged up to 0.1 ppb. When polluted air masses reached the measuring site, concentrations were from 0.2 ppb upward; values above 1 ppb were recorded occasionally [156, 157]. At remote sites such as the equatorial Pacific, HNO_3 concentrations ranged between 0.014 and 0.068 ppb [158]. Data on HNO_3 in ambient air reported in Europe are scarce. Klockow [159] and Slanina [160] measured HNO_3 in Dortmund as well as in Petten and Delft (Netherlands) using the denuder technique. With the exception of one day when 10 ppb were measured, the nitric acid concentrations were generally very low, about one or two ppb.

As HNO_3 is the thermodynamically most stable oxidized nitrogen compound, it is the final fate of most of the atmosperic NO_x. At daytime, the reaction of OH radicals with NO_2 will be the dominating mechanism of converting NO_x to HNO_3. At nighttime, it is probably the rapid reaction of NO_2 with ozone followed by the heterogeneous conversion of NO_3 via N_2O_5 to HNO_3 (see Scheme 28). Thus, the role of clouds and fog in HNO_3 formation needs to be further investigated.

Nitrous Acid (HNO₂)

While the overall mechanism of nitric acid formation and its removal from the atmosphere seems to be reasonably well understood, the origin of HNO_2 can still not be explained statisfactorily. Due to its rapid photolysis, HNO_2 is a rather transient species during the daytime (see Table 1 and Fig. 11). By means of long-path differential UV absorption spectroscopy, HNO_2 was detected in various urban areas of Los Angeles, Cologne und Jülich [161, 162]. Peak levels of up to

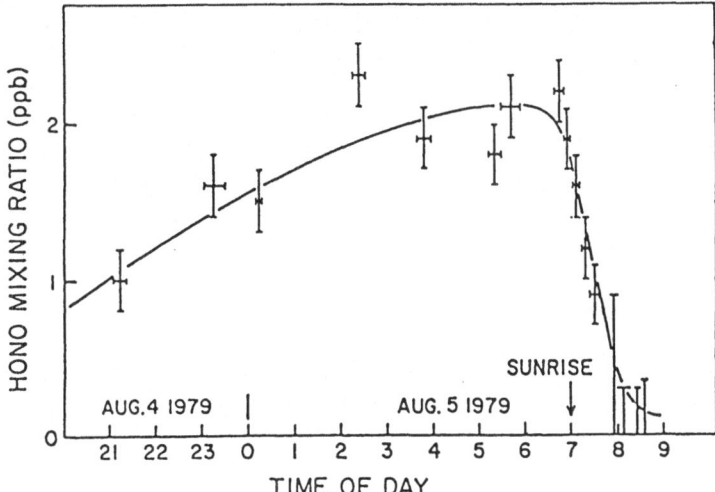

Fig. 11. Concentration-time profile of nitrous acid in Riverside, California [167]

8 ppb were observed in Los Angeles smog [163]. The diurnal concentration profiles of HNO_2 follow the pattern indicated in Fig. 11. After a gradual increase during the night, the HNO_2 concentration decreases rapidly after sunrise. Unlike HNO_3, nitrous acid has been detected only in polluted air. At Deuselbach, a background site in the Federal Republic of Germany, HNO_2 did not exceed 0.05 to 0.1 ppb [161, 162]. No HNO_2 has been detected in a remote area at Loop Hand, Ireland, during the night. In 1972/73, concentrations of HNO_2 between 4 and 21 ppb were measured at a location in southern England. The highest concentrations were found when a trajectory analysis indicated that the air masses had passed over the industrial regions of western Europe [164].

The origin of HNO_2 in polluted air is less clear. Though HNO_2 can be emitted from sources such as car exhaust, power plants and residential heating, chemical formation from precursors must be postulated to account for the concentrations observed in the polluted air. It has been suggested that heterogeneous reactions during the night could be responsible for HNO_2 formation. The homogeneous gas phase reaction:

$$NO + NO_2 + H_2O \rightarrow 2\,HNO_2 \tag{29}$$

is second order in NO_x and thus too slow. A surface-catalysed reaction, however, at deliquescent aerosols could account for the concentrations in polluted atmospheres. Indications of such a surface-catalysed formation of HNO_2 can be taken from the simulation experiments in reaction chambers performed by Chan et al. [165] and Kaiser and Wu [166]. A summary of the formation and fate of HNO_2 and HNO_3 in the atmosphere was published recently by Platt [167].

The most important reaction of HNO_2 in the polluted urban air is its fairly rapid photolysis early in the morning (see Fig. 11). This reaction is the dominant source of OH radicals about one hour after sunrise and can trigger the photosmog process in addition to the reactions of Eqs. 1 and 3. Indeed, model calculations confirm that the formation of photochemical smog can be accelerated by additional OH radical formation via the HNO_2 photolyses [163, 168]. Depending on the concentration of NO_x and hydrocarbons, up to 50% more ozone can be produced in the course of the day if 4% of the initial NO_x is present as HNO_2.

The presence of HNO_2 in polluted urban air can lead to the formation of nitrosamines, a class of notorious cancerogenic compounds. It has been shown that under simulated atmospheric conditions in a smog chamber the photo-oxidation of aliphatic amines yields nitrosamines [169].

Aldehydes and Ketones

The major products of hydrocarbon oxidation in urban smog are aldehydes and ketones, both ubiquitous in metropolitan areas throughout the world. They are formed from most types of hydrocarbons either by reactions with OH radicals (reaction sequence Eqs. 7, 8, 6, 11) or with ozone. The formation of aldehydes through reactions of alkenes with ozone tends to occur later in the day as ozone concentrations increase, reactions of hydrocarbons with OH radicals can occur throughout the daylight hours. In addition, aldehydes and ketones are emitted to the atmosphere as combustion products and by the chemical industry [170]. In general, the concentration of formaldehyde exceeds the concentration of the other aldehydes such as acetaldehyde, propanal, n-butanal and benzaldehyde. Depending on the analytical technique, either total aldehydes or individual aldehydes have been measured. In the early days of photosmog investigations in the 1950s, total aldehydes were generally measured with the bisulfite method. In the County of Los Angeles, the average aldehyde concentrations ranged between 50 and 150 ppb in the 1950s [118]. The diurnal concentration curves for aldehydes are flatter than that of ozone. Although the concentrations of aldehydes decrease at night, the decrease was by a factor of two or less. A compilation of ambient air measurements of formaldehyde and other aldehydes at urban sites in the United States can be found in Ref. [118]. In general, the highest concentrations of aldehydes were measured in the Los Angeles metropolitan area. While the measurements of aldehydes in the United States were made mostly in urban areas, the European measurements were predominantly made at non-urban sites. The reason is partially based on new analytical techniques such as the differential ultraviolet absorption spectroscopy pioneered by Perner and Platt [161, 162] which allows measurement of individual aldehydes such as formaldehyde. While the formaldehyde concentrations in maritime air masses are in the sub-ppb range below 0.2 ppb, the formaldehyde concentration ranged from 0.1 up to 6.5 ppb with the average of about 2 ppb in the area of Jülich between the two cities Aachen and Cologne in West Germany [162, 171]. From July to October 1979, formaldehyde concentrations at a measuring site near Mainz ranged from 0.7 to 5.1 ppb with a mean value of 1.9 ppb [172]. No data on aldehyde concentrations have been reported from urban areas in Europe. Among the aldehydes, the unsaturated aldehydes acrolein and crotonaldehyde are of special interest because of their high eye irritation potential. Though the formaldehyde concentration generally exceeds the sum of concentrations of all the other aldehydes, minor concentrations of acrolein in air can cause annoying effects. In the 1960s, maximum acrolein concentrations between 8 and 14 ppb were recorded in Los Angeles [118]. No acrolein concentrations have been reported for European cities.

Very little is known about the fate of the aldehydes in air. Higher-molecular-weight aldehydes are obviously photo-oxidized to lower-molecular-weight aldehydes. Thus, photo-oxidation of propanal in a smog chamber under photosmog conditions gave acetaldehyde, CO and peroxypropionyl nitrate as the major products [173]. It can be concluded that CO, formaldehyde, acetaldehyde and PAN along with other lower-weight carbonyl compounds are the major products of the photo-oxidation of aldehydes. Little is known about the deposition of aldehydes. There appears to be no rapid mode of dry deposition. Since aldehydes are readily soluble in water there are indications of a wet deposition pathway. Up to ten carbonyl compounds, including formaldehyde, acetaldehyde, propanal, acrolein, higher-molecular-weight aldehydes and benzaldehyde have been identified in fog, mist, cloud and rain water samples collected in Los Angeles [174].

Miscellaneous Compounds

Quite a number of organic compounds have been identified only as products formed in laboratory photo-oxidations of hydrocarbon mixtures of various complexities. The simulation of photochemical smog in reaction cells commonly named "smog chambers" for studying photochemical reactions under simulated urban conditions has led to the identification of a variety of reaction products which might be formed also in ambient air. Smog chambers with solar simulators and advanced analytical techniques such as the Fourier transform infrared spectroscopy have been described in detail in the literature [175–177]. These techniques allow a multi-component in-situ quantitative analysis with high detection sensitivities for measuring concentration-time profiles.

The monoterpenes such as α-pinene and β-pinene are photo-oxidized in the presence of NO_x to form a number of small molecules such as formaldehyde, CO, nitric acid, PAN, etc. and a number of higher-molecular-weight products such as cis-pinonic acid, cis-pinonaldehyde and norpinonic acid from α-pinene and 3-hydroxynopinone, nopinone and 3-ketonopinone from β-pinene [178]. Spicer and Jones [179] reported yields of benzaldehyde, PAN, phthalaldehydes, phenol, cresols, benzoyl alcohol and formaldehyde sufficient to account for 96% of the toluene consumed in the photochemical smog system. Low-boiling products such as formaldehyde, acetaldehyde, glyoxal, methylglyoxal and biacetyl were identified together with high-boiling products such as tolualdehyde, dimethyl phenols and dimethyl nitrobenzene accounting for 62% of the photo-oxidation products of o-xylene [180]. The formation of formic acid under laboratory conditions as well as in ambient air in the Los Angeles area has been reported [147, 181]. The maximum concentration in ambient air was 19 ppb for periods of measurement in 1973 and 1978. There appears to be a growing acceptance that formic acid results from the reaction of formaldehyde with HO_2 radicals [182].

It has been discussed previously that the photochemical oxidation of olefins in air leads to the formation of epoxides, hydroperoxides and peroxides with low molecular weights [183]. Laboratory-scale simulation reactions indeed show the formation and existence of hydroperoxides and peroxyacids during photo-oxidation of low-molecular-weight alkanes [184]. Peroxyacetic acid was found to be an especially stable reaction product. Thus, the smog chamber experiments furnish good reasons to suppose that fairly high yields of low-molecular-weight hydro-peroxides and peroxyacids, in particular peroxyacetic acid, accumulate in polluted urban air.

Long-Range Transport of Photochemical Smog

Wind transports the precursors (NO_x and hydrocarbons) and the reaction products of photochemical smog, and turbulence causes their dispersion within the mixing layer. Variations in wind and turbulence are in turn caused by micro-meteorological factors such as insolation, temperature and their variation in time and space. Since the formation of ozone and other photosmog products in urban air take some time (see Figs. 1, 3, and 10) there is seldom a direct correlation between the ozone maximum concentration and the maximum concentration of its precursors. Only in very calm weather this correlation can be overserved occasionally over fairly small areas. Furthermore, topographical and surface features and their variation in space influence the concentration of local precursor as well as of photosmog products. Thus, transport phenomena remove ozone and other photosmog products under changing micro-meteorological conditions and topographical features from the emission sources of the photosmog precursors. This is one of the reasons why ozone is measured at elevated levels only some time after the primary photochemical processes have started [99].

Various distance and time scales of atmospheric transport are identified with the corresponding scales of meteorological processes, e.g. microscale, mesoscale and synoptic scale. While microscale phenomena describe essentially the dispersion in small urban areas, mesoscale transport may be defined as the transport of pollutants over distances greater than several hundred kilometers and time scales longer than one day. On a local scale, ozone and other photosmog products are transported downwind of an urban area during the daytime, often in the late afternoon with the evening breeze in the closed topography of urban areas at seasides, as in Los Angeles or New York [185]. Regarding transport phenomena on the microscale one has to consider the so-called "Sunday Effect" indicating that Sunday ozone concentrations measured in certain urban areas are similar to those occurring on workdays, despite markedly decreased emissions due to different traffic patterns. There are indications of a recirculation of the photosmog by the sea-breeze in the closed topography of Greater Athens, Greece [103, 186]. It is, however, difficult to assess whether the change in ozone concentration is due to a horizontal transport associated with the sea-breeze frontal passage or to changes in vertical mixing depth.

Over the sea, ozone and its precursors can find particular conditions such as a greater atmospheric stability due to a smaller deposition velocity [187, 188], limited vertical exchange and lower destruction processes and thus can be transported over fairly long distances without undergoing dilution or transformation. Because of its low destruction rate over water, ozone has a lifetime 15 times longer over the sea than on dry land surfaces [187]. Thus, on the microscale, ozone transport from urban areas can be best monitored on a remote island [189].

Many observations indicate that urban and rural ozone levels may be strongly coupled to each other. The influence of urban traffic on ozone pollution in adjacent rural areas has been demonstrated, e.g. for New York [195, 196] where it is independent of the day in the week. Thus, the conclusion is that, as in the Los Angeles area, it is the rural or suburb areas downwind of the primary pollutants which experience the higher ozone levels. Mesoscale transport spans the region between local and long-range transport. Such effects as land and sea-breeze circulations might be operative on this scale. Complex terrain influence, such as valley-mountain circulations and flow over mountains ranges, fall into the mesoscale transport regime. There is no general agreement about the time when mesoscale transport extends to long-range transport. Photochemical smog is typically a mesoscale phenomenon with transport distances of many hundreds of kilometers. Recently several authors pointed out that the transport of ozone [161, 190–192] and PAN [128] can occur on trajectories of many hundreds of kilometers length. Though PAN is genereally considered by chemists as an unstable compound, it obviously can have a long atmospheric lifetime when diluted in air [13]. Since the absorption cross-section of PAN leads to a photodissociation lifetime of approximately four months in the troposphere [193] and since the PAN lifetime for the reaction with OH radicals according to Eq. 20 is long [194], all prerequisites of a long-range transport are fulfilled. Thus, PAN is a useful indicator of photochemically polluted air. The comparison of PAN levels in Denmark and southern Sweden with those observed under optimum conditions for PAN formation from local sources indicated that long-range transport contributes substantially to the presence of PAN in southern Scandinavia [128]. A detailed discussion of selected episodes of long-range transport of photo-oxidants to Scandinavia can be found in Refs. [91] and [93].

Normally, the long-range transport of ozone, PAN and other photo-oxidation products occurs during the nighttime above the nocturnal inversion where the air parcel being transported is isolated from the effects which remove the air pollutants. While high nocturnal levels of ozone are observed in the mixing layer, the concentration at ground levels are usually very low due to depletion. Venting of the air pollutants from within to above the mixing layer by cumulus clouds may be another mechanism that leads to long-range transport of the photo-oxidation products. Although cloud-vented pollutants are transported by a wind environment differing from that of the pollutants within the mixing layer, they could eventually reenter the mixing layer to affect the surface layer concentrations of the atmospheric pollutants. This again illustrates the importance of monitoring atmospheric pollutants at different heights in order to be able to evaluate the level of air pollution, in particular that of the secondary air pollutants.

Tracing ozone-rich air masses back to specific urban plumes is often difficult. In general, the air mass is modified continuously due to mixing with pollutant emissions from ground sources and mixing with other air parcels. One method of analysing long-range transport consists in analysing daily ozone isopleth maps and comparing them to meteorological maps and air parcel trajectoris [197–199]. It is even more effective to combine the air-mass-trajectory model with aircraft sampling [200, 201]. Another approach is modeling the long-range transport of ozone and other oxidants in such a manner that the modelled air mass follows the atmospheric mean motion. A model which transports ozone and other oxidants over long distances of about 1,000 km from the European continent to the British Isles has been described by Hov et al. [191]. Indeed, during the very hot summer of 1976 in Central Europe, the analysis of the air mass trajectories indicated the long-range transport of ozone from Central Europe to southeast England [86, 88]. Transport phenomena of photo-oxidants in the Rhine Valley between Cologne and Bonn have been observed [77].

Mathematical Models Describing Long-Range Transport

It was outlined in the section on computer modelling of photochemical smog that stationary photosmog models are connected with transport models. Though it is beyond the scope of this chapter to discuss the long-range transport models in detail, a brief account of the essential features will be given here.

In pollution-transport modelling, oxidants are subjected to a variety of physical and chemical processes that transport, dilute, transform and remove the oxidants. Atmospheric parameters which are taken into account during these processes include variable wind speed, surface roughness, mixing height and diffusivity. These processes are generally described mathematically by a set of coupled partial nonlinear differential equations in four dimensions. The change of concentration in time and space of each oxidant, i, must satisfy the atmospheric diffusion equation [202, 203]:

$$\frac{\partial c_i}{dt} = -\left(u\frac{\partial c_i}{\partial x} + v\frac{\partial c_i}{\partial y} + w\frac{\partial c_i}{\partial z} \right)$$
$$+ \frac{\partial}{\partial x}\left(k_H \frac{\partial \bar{c}_i}{\partial x} \right) + \frac{\partial}{\partial y}\left(k_H \frac{\partial \bar{c}_i}{\partial y} \right) + \frac{\partial}{\partial z}\left(k_v \frac{\partial \bar{c}_i}{\partial z} \right) \qquad (30)$$
$$+ R_i$$
$$+ P_i.$$

The first term on the right represents the mean advection (horizontal and vertical transports by the wind) with u, v and w being the mean values of the wind velocity in the x, y, z dimensions; the second term comprises the turbulent diffusion of oxidants in air; R_i is the reaction rate of pollutant i, and P_i summarizes the rate for wet and dry depositions. k_H and k_v are the horizontal and vertical coefficients for turbulent diffusion derived by the so-called k-theory [204]. The transport by eddy motions is usually parametrized as the scalar product of a diffusion tensor and the gradient of the mean pollutant concentration. Thus, the driving forces for atmospheric transport and dispersion, such as the wind speed and wind direction, enter the atmospheric diffusion equation Eq. 30 through the component variables u, v and w and the turbulent diffusivities k_H and k_v. Since the data have to be known as functions of space and time, which are often not available for a given area, these parameters are estimated or calculated from measured meteorological data from synoptic weather stations. The boundary condition is given by the inversion height. Generally, Eq. 30 contains a term describing the rate of emission of a pollutant i from a source. For the transport of photochemical oxidants, however, it can be assumed that they are homogeneously dispersed after completion of the photosmog cycle in a given airshed modelling system. During the transport a two-dimensional model, which follows a parcel of air as it travels, is often appropriate if vertical mixing of the pollutants is sufficiently rapid. The main advantage of a two-dimensional model consists in the elimination of the third spatial dimension. This implies a substantial reduction in computation time and computer storage requirements. Furthermore, it allows to incorporate more detailed chemical and photochemical processes in the model. Such two-dimensional models, in which the air parcel traverses along the center line of an area source, are based on a Lagrangian or Eulerian grid approach. In such trajectory models the co-ordinate system is advected by the horizontal wind field.

Box Models

The simplest area source model is the so-called single box model. The pollutants are assumed to be completely mixed within a single box which covers the area under investigation and extends upward to the mixing height. The time dependence of the concentration of a pollutant i is given by:

$$\frac{dc_i}{dt} = \frac{(c_i - c_0)}{H} \frac{dH}{dt} - (c_i - c_0)\frac{\bar{u}}{L} + R_i, \tag{31}$$

where c_0 is the background concentration outside the box, H the height, L the width of the box, and u the mean wind speed used to calculate the flux of pollutants through the box. No turbulence parameters are needed in this model because it works with a time-averaged meteorology. Hanna [205] has described a simple dispersion model for the analysis of pollutants in the Los Angeles area. It is not surprising that these simple box models give contradictory results. While Venkatram [206] did not find an agreement between predicted and measured pollutant concentrations, Shreffler and Schere [207] recommend photochemical box models for urban air quality simulations. A photochemical box model for the analysis of photochemical smog episodes in the Cologne-Bonn area was developed by Scherer and Stern [208]. A box model with 300 reactions in a huge box of 450×360 km^2 for calculating a seven-day photosmog episode in south England has been described in detail by Derwent and Hov [45]. A photochemical box model for urban air with 143 chemical reactions has been published by Graedel et al. [209].

Eulerian Grid Models

If the concentration distribution of pollutants has to be calculated over a large area, an array of single boxes is used. For every box the complete mass balance has to be computed. The grid models call for detailed input data. Since the meteorological transport model involves a number of computational boxes, the chemical reactions are often described by lumped mechanisms for selected classes of chemical compounds such as olefins, alkenes, aromatics aldehydes, etc. The numerical solutions are obtained at all horizontal and vertical grid points. In general, grid models can be used for the prediction of the photochemical oxidant concentration in a given area. Grid models have been described in detail by Reynolds et al. [210], McCracken et al. [211] and Schiavone and Graedel [212]. One of the most advanced grid models is the so-called SAI-model of Reynolds et al. [210].

Langrangian Models

These so-called trajectory models predict the changes of pollutant concentrations along a chosen wind trajectory. In general, the advective term of the transport equation Eq. 30 is omitted because the assumption is made that the vertical advective transport is small in comparison to turbulent diffusion. If, however, the meteorological conditions in the area are such that the vertical component of the wind field is large, the advective transport term must be retained in Eq. 30. The vertical integrity of the moving air parcel can only be maintained if the effects of wind shear can be neglected. This is a critical assumption and a major source of error in trajectory model calculations, especially those which involve long transport times. A quantitative assessment of the simplifications in trajectory models is presented by Liu and Seinfeld [213]. Noted trajectory models for photochemical oxidants are those of Eschenroeder and Martinez [46] and Hov et al. [191]. A detailed assessment of mathematical modelling of photochemical air pollution can be found in Ref. [214] and a critical evaluation of the kinetic and mechanistic data for photochemical smog modelling in Ref. [215]. A recent account on the long-range transport models for photochemical oxidants and their precursors was presented during a jointly organized international conference by U.S. EPA and OECD [216].

Control of Photochemical Smog

When discussing the measures of control of photochemical oxidants, one has to consider the damage to the human population, animals, vegetation and to materials. Since ozone is the main secondary pollutant of photochemical smog, in general the human symptoms and vegetation injuries caused by ozone is considered. Again, it is beyond the scope of this chapter to discuss these important problems in detail, and only a brief account of this topic will be given here.

Ambient Air Quality Criteria for Ozone

The difficulty in setting proper air quality criteria for ozone is that ozone is a natural trace gas of the troposphere with average concentrations of 20–40 ppb [58]. All the assessment studies indicate that health effects and, in particular, damage to the vegetation occur close to levels of the natural background concentration of ozone. Table 7 summarizes the effects of ozone, which led to the ambient air quality criteria for ozone.

Table 7. Conversion of effects of ozone to standards [217]

Concentration of ozone in air (ppm)	Symptoms in humans, and threshold of damage to vegetation
1.0	Cough; extreme fatigue; lack of coordination; increased airway resistance; decreased forced expiratory volume.
0.5	Chest constriction; impaired carbon monoxide diffusion capacity; decrease in lung function without exercise.
0.3	Headache; chest discomfort sufficient to prevent completion of exercise; decrease in lung function in exercising subjects.
0.25	Increase in incidence and severity of asthma attacks; moderate eye irritation.
0.15	In sensitive individuals, reduction in pulmonary lung function; chest discomfort; irritation of the respiratory tract, coughing and wheezing. Threshold for injury to vegetation.
0.12	United States National Primary and Secondary Ambient Air Quality Standard, attained when the expected number of days per calendar year with maximum hourly average concentrations above 0.12 ppm is equal to or less than 1, as determined in a specified manner.

The toxicological experiments indicate that higher than ambient doses of ozone may have an impact on the biochemical activity of the blood, reduce pulmonary function, and increase general respiratory discomfort. Though no substantial relationship between ozone concentrations and human mortality has been found, stronger associations have linked increased ozone levels with eye irritation, coughs, chest discomfort, and fatigue [217]. Ozone damages the membranes surrounding pulmonary capillaries and water passes from blood vessels to lung alveoli causing what is termed pulmonary edema. A recent review summarizes the impact of ozone and PAN on humans and animals [218]. The total annual health costs attributable to photochemical oxidants for the period 1971–1973 were estimated to range from $ 120 to over $ 240 million in the U.S. [219]. Compared to human beings, trees and plants are obviously even more susceptible to ozone concentrations higher than the natural background concentration. The effects of ozone in plants are often proportional to dosage (concentration x exposure time), i.e. there is a no-threshold dose response. Concentrations of twice the natural background concentration of ozone can produce visible foilar symtpoms and reductions in growth or yield. Some vegetables such as head lettuce and spinach are particularly sensitive to ozone. Estimated production losses in major crops due to ozone at a seasonal 7-hour/day mean ozone concentration of 40–60 ppb, compared to a base concentration of 25-ppb, were: head lettuce (53–56%), peanut (14–17%), soybean (10%) and turnip (7%) [220]. An estimate of direct crop losses assuming that all areas of the U.S. meet the U.S. Air Quality Standard for ozone of 120 ppb, showed a loss of 2–4% of the crop production which is equal to a loss between one and two billion dollars [220]. A summary of the effects of photochemical oxidants on vegetation has been published by Jacobson [221] and Guderian et al. [222]. It has been argued that one of the multiple stress diseases responsible for the dying forests (Waldsterben) in Europe is caused by photochemical oxidants. Thus, international activities in research and legislation will be necessary in Europe if the problem of harmful effects of photochemical oxidants on vegetation is not to increase in the coming decade.

Besides its damaging effect on humans and vegetation, ozone may also cause damage to fibres, dyes, paints and other organic materials. In fact, natural rubber and synthetic polymers suffer from a deterioration of the mechanical structure by the action of ozone. Ozone reacts with double bonds in organic material. In this process, the material cracks and becomes brittle. Regarding the degradation of rubber and other elastomers by ozone, the prevention or retardation of this process by means of antiozonant additives is fo great interest. Photochemical oxidants also damage exterior household paints [223] and pigments and, consequently, are a major hazard to works of art [224]. The effect of photochemical oxidants on materials is reviewed by Schreiber [225].

National ambient air quality standards for photochemical oxidants were first established in the U.S. in 1971 [17]. This value, measured as ozone, was initially set at an hourly average concentration not to exceed 80 ppb more than once per year. At that time, ozone concentrations in rural and remote area were believed to be low relative to the standard. Furthermore, the U.S. Environmental Protection Agency (EPA) decided to refer the health and welfare standard for photochemical air pollution exclusively on ozone, rather than on total oxidants. Since air quality standards generally contain a built-in safety factor, which is on the order of 3–5 for health effects, the value of 80 ppb was in line with common practice. However, subsequent studies of the ozone concentration in rural areas as well as on transport phenomena throughout the world, as described in the preceding sections, have shown that ozone is almost unique as an air pollutant, because the line between harmless and harmful is very narrow, probably more so for ozone than for any other air pollutant. In response to these findings, the EPA promulgated in 1979 a revised ambient air quality standard for ozone stating that the expected number of days per calendar year with daily maximum ozone concentrations exceeding 120 ppb must be one. Some air quality goals and standards for ozone have now been established in several countries (Table 8).

The German value is a value recommended by VDI (Verein Deutscher Ingenieure) which considers the effect of ozone on human beings only [226]. Since the natural background concentration of ozone is close to levels where adverse effects, especially on vegetation, have been documented, well-balanced fixing of an

Table 8. Hourly ambient air quality standards or guidelines for ozone in several countries

Country	Concentration [ppb]	Frequency	Status
USA	120	Once a year	Legal standard
Germany[a]	100		Proposed guideline
Japan	60	Once a year	Legal standard
Sweden	60	One a month	Proposed guideline
Canada	80		Maximum acceptable level
WHO[b]	60		Recommended value

[a] VDI-Guideline 2310, August 1984.
[b] World Health Organisation.

ambient air quality standard for ozone is difficult and an economic problem. Fixing this value at a too high or too low level can cause severe economic losses. These considerations do not include the deleterious effects of PAN, a well-known phytotoxicant. Since, however, the diurnal cycle of the concentration of PAN correlates with that of ozone [87, 124] (see Fig. 10), the control of ozone as an indicator of photochemical air pollution appears to be a good policy.

Air Pollution Control Principles

Unlike for common primary air pollutants such as SO_2, fixing of air quality standards for secondary air pollutants such as ozone is extremely difficult because standards have to be established for the precursor pollutants NO_x and the NMHC (non-methane hydrocarbons). The crucial question is how much reduction of both of the precursors is needed to ensure that ozone concentrations in excess of the ambient air standard will not be formed? Due to the complex system of long-chain free radical reactions in the photochemical smog cycle, a simple linear reduction of NO_x and NMHC does not result in the same reduction in ozone or photochemical oxidants, in other words, the effects of reducing the precursor concentrations is a nonlinear problem. Nevertheless, control strategies for photochemical oxidants should consist in reducing the NO_x/NMHC ratio. Furthermore, it is important to consider the ambient air quality standard of ozone for an area which one wants to protect. On a local scale it is often sufficient to reduce only the NMHC.

Unilateral reduction of NMHC levels then leads to increasing ozone levels in downwind areas of a city, where the reacting air masses need longer reaction times to build up photochemical smog. Thus, on the local scale, NMHC control only delays the production of photochemical oxidants because the conversion of NO to NO_2 is slowed down over the city. Outside of the urban areas, however, the precursors together with further infusions of organic vapours create a greater potential for photochemical smog. Another factor makes forecasting of peak ozone levels difficult, namely the dependency of secondary air pollutant concentrations on meteorological parameters.

Nevertheless, for control strategies in urban areas the following independent measures have been proved to be successful in the reduction of photochemical air pollution:

Emission control of motor vehicles and stationary sources. This includes the modification in operation of conventional combustion engines, reduction and exhaust emission control of NO_x and hydrocarbons in connection with catalytic converters, smoothening the traffic flow in an urban area which leads to a reduction of the frequently idling and stop-and-go traffic as well as removal of gaseous pollutants from industrial effluent streams. The last point should include also the removal of particulate matter from effluent gases because there are indications of a heterogeneous photocatalytic effect of particulate matter on photochemical smog formation [227] and the degradation of other atmospheric pollutants [228].

Societal options. Particularly in smoggy weather situations, societal options such as staggered working hours, car pooling, restricted traffic in congested area as well as increased public transportation can lower substantially the emission of NO_x and hydrocarbons.

Regional planning of monitoring sites. It is no trivial task to install a network of monitoring stations. Precursors and photochemical oxidants have to be measured in such a manner that they are representative of a given area. This includes, for instance, that ozone is not measured near highways or on the downwind side of power plants. In this context, the question arises how many stations are required to estimate the regional ozone design value [229]. If possible, vertical ozone profiles should be measured.

Forecasting methods. Based on measured data of the precursors of photochemical smog, NO_x and NMHC, the isopleth modelling approach can predict ozone concentrations for each cell in a network of cells that overlie the area of interest. The isopleth methodology was discussed in detail in a previous section of this article. It is important to note that the interpolation scheme of isopleths (see Fig. 6) is not linear. Having in mind that the isopleth method was developed in smog chamber experiments where no meteorology is involved, the quality of forecasting depends on meteorological parameters. In the "closed" topography of an urban area, however, it is occasionally possible to predict the ozone concentration at 1 p.m. from the NO_x and NMHC concentrations measured between 6 and 9 a.m. [39, 230]. According to Kinosian [231] who used the so-called EKMA isopleths, the maximum ozone concentration depends linearly on the mean concentration of the precursors. An alternative to the so-called EKMA-method has been suggested by Jones et al. [232]. A simple model for the prediction of ozone in urban areas based on the Carbon-Bond mechanism was recently published by Laird et al. [233] and a model based on the so-called Box-Jenkins technique by Simpson and Layton [234].

A purely empirical approach of predicting ozone concentrations from measured precursor concentrations is to plot the mean precursor concentration of a given area versus the mean ozone concentrations measured in the same area. On the basis of a vast amount of historical data for that particular area it is often possible to predict the afternoon mean ozone concentration from mean precursor concentrations measured early in the morning. This simple method, however, yields useful data only if the urban area under consideration is fairly isolated from other precursor emission sources and if an extended network of measuring stations is available. Under these circumstances the local meteorology and the topography are "integrated" in the local network of measuring sites.

In general, it is not obvious that a control stategy developed for one local area will contribute to a reduction in photochemical oxidants in another area or on the downwind side.

Chemical Control of Photochemical Smog

Although strategies for photochemical smog control have been focussed almost exclusively on reducing the emission of the precursors NO_x and reactive hydrocarbons, the possiblity of suppressing photochemical smog by spraying into the lower atmosphere chemicals which are known to be free radical traps or so-called smog inhibitors has been discussed for at least 20 years [235]. The basic idea is that the photochemical smog reactions are long-chain free radical processes which can be retarded by adding free radical scavengers to the atmosphere early in the morning before photochemical processes begin. The question of freshening polluted urban air by treatment with photosmog inhibitors has reattained topicality in the 1970s, largely on account of the diethylhydroxylamine (DEHA) controversy [236, 237]. Jayanti et al. [238], after exhaustive qualitative studies on this and several other aliphatic amines as smog inhibitors in smog chambers, suggested it to become an object for field tests. The idea is to introduce the scavenger into the urban air to slow down the conversion of NO to NO_2 from the usual 2–6 h to 6–12 h so that the sun will be gone and the air will be cleaned in the evening and overnight by normal dispersion. The chemical method using DEHA is projected to be cheaper by at least a factor of 100 than automobile control devices and could be used only when and where it is needed [236]. On the basis of quantitative experiments measuring the reaction rate constants of DEHA with the chain-carrying OH radical [239] and the $O[^3P]$ atom [240], Filby and Güsten [240, 241] have shown that at least for suppressing the reaction with $O[^3P]$ at the early stage of the photochemical smog process, DEHA concentrations in the low ppm range would be necessary. Though DEHA is not mutagenic in concentrations smaller than 2 ppm [242], DEHA's odour threshold level of 0.5 ppm prohibits its introduction into polluted urban air at concentrations higher than 0.5 ppm.

Final Remarks

Since there is little difference in the solar intensities integrated over daylight hours in Los Angeles (35°N latitude), generally regarded as photosmog city number one, and Central Europe (50°N latitude, Paris, Frankfurt) during the summer solstice (May until September), the potential of photochemical air pollution in Europe in high-pressure situations with anticyclonic weather conditions is obvious. During the winter, however, due to a strong decrease in solar intensity and a limited number of cloudless days, the potential for photochemical air pollution is smaller by a factor of about ten [243]. Furthermore, contrary to the emission of SO_2, the emissions of the photosmog precursors NO_x and hydrocarbons have strongly increased over the last 15 years. Thus, the formation of photochemical oxidants in the lower atmosphere of European cities and metropolitan areas is a growing menace. Due to the density of industrial and urban areas in Europe and the phenomenon of long-range transport of photochemical oxidants, local control strategies are condemned to yield but limited success. There is a definite need for a joint European effort to develop control strategies for the large-scale formation of photochemical oxidants, in particular if one considers the suspicion that photochemical oxidants play a role in the dying forest phenomenon (Waldsterben). There is a need for a joint monitoring programme in Europe of the precursors as well as of the photochemical oxidants such as ozone and PAN. Ozone, as an indicator of photochemical smog, should be monitored continuously at ground level stations as well as by aircraft. In addition to supporting the preparation of emission inventories, the development of large-scale photochemical models on the relevant time and spatial scale of European areas should be funded in order to assess the origins of polluted air masses. A programme for continuous measurements of ozone damage to sensitive bioindicators such as tobacco plants [244] should complement these European efforts.

The knowledge of and experience in photochemical air pollution gathered in the U.S., in particular in California, must be used today in Europe if the problem is not to increase in the future.

Acknowledgements

The author wants to thank those who have provided him with pre- and reprints on the topic of photochemical air pollution. The magnitude of the literature prevented a comprehensive citation of all those who have contributed to the work in this area over the last 15 years. The author must apologize to his colleagues whose work has been unintentionally not cited.

References

1. Leighton, P.A.: Photochemistry of Air Pollution, Academic Press, New York, 1961
2. Altshuller, A.P., Bufalini, J.J.: Photochemical Aspects of Air Pollution: A Review, Environ. Sci. Technol. 5, 39–64 (1971)
3. Demerjian, K.L., Kerr, J.A., Calvert, J.G.: The Mechanism of Photochemical Smog, Advan. Environ. Sci. Technol. 4, 1–256 (1974)
4. Niki, H., Daby, E.E., Weinstock, B.: Mechanisms of Smog Reactions, Chapter 2, in "Photochemical Smog and Ozone Reactions", Advan. Chem. Series 113, 16–57 (1972)
5. Güsten, H., Penzhorn, R.-D.: Photochemische Reaktionen atmosphärischer Schadstoffe, Naturwiss. Rundschau 27, 56–68 (1974)
6. Haagen-Smit, A.J.: Ind. Engng. Chem. 44, 1342 (1952)
7. Haagen-Smit, A.J., Fox, M.M.: Ind. Engng. Chem. 48, 1484 (1956)
8. Chuck, E.A., Ford, H.W., Stephens, E.R.: Air Pollution effects of irradiated automobile exhaust as related to fuel composition. Rept. No. 26, Air Pollution Foundation, San Marino, Ca. (1958)
9. Doyle, G.J., Renzetti, N.A.: J. Air. Pollut. Control. Assoc. 8, 23 (1958)
10. Middleton, J.T., Haagen-Smit, A.J.: J. Air. Pollut. Control. Assoc. 11, 125 (1961)
11. Middleton, J.T.: Ann. Rev. Plant. Physiol. 12, 431 (1961)
12. "Air Quality Criteria for Photochemical Oxidants", U.S. Department of Health, Education, and Welfare, Public Health Service, Environmental Health Service, National Air Pollution Control Administration, Washington, D.C., March, 1970; NAPCA publication No. AP-63
13. Stephens, E.R.: The Formation, Reactions, and Properties of Peroxyacetyl Nitrates in Photochemical Air Pollution, Advan. Environ. Sci. Technol. 1, 119–146 (1969)
14. Odaira, T.: Photochemical Smog in Tokyo. The Tokyo Metropolitan Research Institute for Environmental Protection, Shinkodo Co., Ltd., November 1972
15. Hawke, G.S., Iverbach, D.: Atmos. Environ. 8, 597 (1974)
16. Cox, R.A., Eggleton, A.E.J., Derwent, R.G, Lovelock, J.E., Pack, D.H.: Nature 255, 118 (1975)
17. "National primary and secondary ambient air quality standards", Federal Register 36, 8186 (1971)

18. World Health Organization "Photochemical oxidants", Environmental Health Criteria No. 7, Geneva, Switzerland, 1978
19. Calvert, J.G.: Test of the Theory of Ozone Generation in Los Angeles Atmosphere, Environ. Sci. Technol. *10*, 248–256 (1976)
20. Jones, I.T.N., Bayes, K.D.: J. Chem. Phys. *59*, 4836 (1973)
21. Gaedtke, H., Troe, J.: Ber. Bunsenges. Phys. Chem. *79*, 184 (1975)
22. Jackson, J.O., Stedman, D.H., Smith, R.G., Hecker, L.H., Warner, P.O.: Rev. Sci. Instrum. *46*, 376 (1975)
23. Zafonte, L., Rieger, P.L., Holmes, J.R.: Environ. Sci. Technol. *11*, 483 (1977)
24. Harvey, R.B., Stedman, D.H., Chameides, W.: J. Air Pollut. Control Assoc. *27*, 663 (1977)
25. Bahe, F.C., Schurath, U., Becker, K.H.: Atmos. Environ. *14*, 711 (1980)
26. Dickerson, R.R., Stedman, D.H.: Environ. Sci. Technol. 14, 1261 (1980)
27. Madronich, S., Hastie, D.R., Ridley, B.A., Schiff, H.I.: J. Atmos. Chem. *1*, 3 (1983)
28. Parrish, D.D., Murphy, P.C., Albritton, D.L., Fehsenfeld, F.C.: Atmos. Environ. *17*, 1365 (1983)
29. Calvert, J.G.: The Chemistry of the Polluted Troposphere, in: Chemistry of the Unpolluted and Polluted Atmosphere, pp. 425–456, H.W. Georgii and W. Jaeschke (eds.), D. Reidel Publ. Comp., Dordrecht-Boston-London, 1982
30. Finlayson-Pitts, B.J., Pitts, J.N. Jr.: The Chemical Basis of Air Quality: Kinetics and Mechanisms of Photochemical Air Pollution and Application to Control Strategies, Advan. Environ. Sci. Technol. *7*, 75–162 (1977)
31. Hampson, R.F., Braun, W., Brown, R.W., Garvin, D., Herron, J.T. Huie, R.E., Curylo, M.J., Laufer, A.H., McKinley, J.D., Okabe, H., Scheer, M.D., Tsang, W., Stedman D.H.: Survey of Photochemical and Rate Date for Twenty-eight Reactions of Interest in Atmospheric Chemistry, J. Phys. Chem. Ref. Data. *2*, 267–312 (1973)
32. Pitts, J.N. Jr.: Photochemical Air Pollution: Singlet Molecular Oxygen as an Environmental Oxidant. Advan. Environ. Sci. Technol. *1*, 289–337 (1969)
33. Penzhorn, R.D., Güsten, H., Schurath, U., Becker, K.H.: Environ. Sci. Technol. *8*, 907 (1974)
34. Penzhorn, R.D., Güsten, H., Schurath, U., Becker, K.H.: Staub-Reinhalt. Luft *35*, 95 (1975)
35. Simonaitis, R., Heicklen J.: J. Phys. Chem. *77*, 1096 (1973)
36. Peterson, J.T.: Calculated Actinic Fluxes (290–700 nm) for Air Pollution Photochemistry Applications. EPA-Report 600/4-76-025 (1976)
37. Calvert, J.G.: The Homogeneous Chemistry of Formaldehyde Generation and Destruction within the Atmosphere, in: Proceedings on the NATO Advanced Study Institute on Atmospheric Ozone: Its Variation and Human Influences, M. Nicolet, A.C. Aikin (eds.), Report No. FAA-EE-80-20, pp. 153–190 (1980)
38. Becker, K.H., Löbel J., Schurath, U.: Bildung, Transport und Kontrolle von Photoxidatien, in: Luftqualitätskriterien für photochemische Oxidantien, Report 5/83, pp. 3–132, Umweltbundesamt, E. Schmidt Verlag, Berlin, 1983
39. Dimitriades, B.: Environ. Sci. Technol. *11*, 80 (1977)
40. Becker, K.H., Löbel, J., Schurath, U.: Staub-Reinhalt. Luft *38*, 278 (1978)
41. Sakamaki, F., Okuda, M., Akimoto, H., Yamasaki, H.: Environ. Sci. Technol. *16*, 45 (1982)
42. Laub, R.J., Smith, C.A., cited in [29], Ref. 4
43. Bruce, W.G., Jr., Hanst, P.L., Bufalini, J.J., Noonan, R.C.: Environ. Sci. Technol. *10*, 58 (1976)
44. Hov, Ø., Isaksen, I.S., Hesstvedt, E.: Atmos. Environ. *12*, 2469 (1978)
45. Derwent, R.G., Hov, Ø.: Environ. Sci. Technol. *14*, 1360 (1980)
46. Eschenroeder, A.Q., Martinez, J.R.: Concepts and Applications of Photochemical Smog Models, Advan. Chem. Ser. *113*, 101–168 (1972)
47. Hecht, T.A., Seinfeld, J.H., Dodge, M.C.: Environ. Sci. Technol. *8*, 327 (1974)
48. Falls, A.H., McRae, G.J., Seinfeld, J.H.: Int. J. Chem. Kinet. *11*, 1137 (1979)
49. Whitten, G.Z., Hogo, H., Killus, J.P.: Environ. Sci. Technol. *14*, 690 (1980)

50. Schurath, U., Goeth, N., Henrich, K., Lippmann, H.: Untersuchungen zur Entwicklung eines chemischen Reaktionsmodells atmosphärischer Spurengasumsetzungen. Teil II. Analyse des chemischen Reaktionsmechanismus im "SAI Airshed Model 1978". Report No. 81-10402511/02, Umweltbundesamt, Berlin, 1982

51. Hübler, G., Perner, D., Platt, U., Tönnissen, A., Ehhalt, D.H.: J. Geophys. Res. *89*, 1309 (1984)

52. Atkinson, R., Darnall, K.R., Lloyd, A.C., Winer, A.M., Pitts, J.N. Jr.: Kinetics and Mechanisms of the Reaction of the Hydroxyl Radical with Organic Compounds in the Gas Phase, Advan. Photochem. *11*, 375–487 (1979)

53. Güsten, H., Filby, W.G., Schoof, S.: Atmos. Environ. *15*, 1763 (1981)

54. Farhataziz, Ross, A.B.: Selected Specific Rates of Reactions of Transients from Water in Aqueous Solutions. III. Hydroxyl Radical and Perhydroxyl Radical and Their Radical Ions. Report NSRDS-NBS 59, 1–113 (1977)

55. Zetzsch, C.: Predicting the abiotic degradability of organic chemicals in the atmosphere by OH using structure reactivity relations. Conference on "Chemicals in the Environment", pp. 302–312, Copenhagen, 18–20 Oct. 1982

56. Güsten, H., Klasinc, L., Marić, D.: J. Atmos. Chem. *2*, 83 (1984)

57. Crutzen, P.J.: The global distribution of hydroxyl, in: Atmospheric Chemistry, E.O. Goldberg (ed.), pp. 313–328, Springer-Verlag, Berlin-Heidelberg-New York 1982

58. Singh, H.B., Ludwig, F.L., Johnson, W.B.: Atmos. Environ *12*, 2185 (1978)

59. Johnson, W.B., Viezee, W.: Atmos. Environ. *15*, 1309 (1981)

60. Kley, D., Drummond, J.W., McFarland, M., Liu, S.C.: J. Geophys. Res. *86c*, 3153 (1981)

61. Seiler, W., Fishman, J.: J. Geophys. Res. *86*, 7255 (1981)

62. Fishman, J., Solomon, S., Crutzen, P.J.: Tellus *31*, 432 (1979)

63. Liu, S.C., Kley, D., McFarland, M., Mahlman, J.D., Levy II, H.: Geophys. Res. *85*, 7546 (1980)

64. Logan, J.A., Prather, M.J., Wolfsy, S.C., McElroy, M.B.: Tropospheric Chemistry: a global perspective. J. Geophys. Res. *86*, 7210–7254 (1981)

65. Hameed, S., Stewart, R.W.: J. Geophys. Res. *88*, 5153 (1983)

66. Regener, V.H.: Z. Physik *109*, 642 (1938)

67. Ehmert, A.: J. Atmos. Terrest. Phys. *2*, 189 (1952)

68. Junge, C.E.: Tellus *14*, 363 (1962)

69. Jost, D.: Pure Appl. Chem. 24, 643 (1970)

70. Lahmann, E.: Umschau Wiss. Techn. *21*, 693 (1969)

71. Junge, C. in: Umschau *21*, 704 (1969)

72. Guicherit, R., Jeltes, R., Lindqvist, F.: Environ. Pollut. *3*, 91 (1972)

73. Atkins, D.H.F., Cox, R.A., Eggleton, A.E.J.: Nature *235*, 372 (1972)

74. Derwent, R.G., Stewart, H.N.M.: Nature *241*, 342 (1973)

75. Becker, K.H., Schurath, U.: Staub-Reinhalt. Luft *35*, 156 (1975)

76. Becker, K.H., Schurath, U., Georgii, H.W., Deimel, M.: Untersuchungen über Smogbildung, insbesondere über die Ausbildung von Oxidantien als Folge der Luftverunreinigung in der Bundesrepublik Deutschland, Research Report 79-10402502/03/04, Umweltbundesamt, August 1979

77. Georgii, H.W., Fricke, W., Rudolf, W., Deimel, M., Becker, K.H., Schurath, U.: VDI-Berichte *270*, 19–24 (1977)

78. Birkle, M.: Staub-Reinhalt. Luft *35*, 41 (1975)

79. Oberländer, W., Siegel, D.: VDI-Berichte *270*, 57–61 (1977)

80. Edinger, J.G.: Environ. Sci. Technol. *7*, 247 (1973)

81. Gloria, H.R., Bradburn, G., Reinisch, R.F., Pitts, J.N. Jr., Behar, J.V., Zafonte, L.: J. Air Pollut. Control. Assoc. *24*, 645 (1974)

82. Fricke, W.: Die Bildung und Verteilung von anthropogenem Ozon in der unteren Troposphäre. Bericht des Instituts für Meteorologie und Geophysik der Universität Frankfurt, Nr. 44, pp. 1–133 (1980)

83. Neuber, E., Georgii, H.W., Müller, J.: Physico-Chemical Behaviour of Atmospheric Pollutants, Proceedings of the Second European Symposium, 29. Sept.–1. Oct. 1981, Versino, B. and Ott, H. (eds.), p. 469, D. Reidel Publ. Company, Dordrecht (Netherlands), 1982

84. Van Duuren, H., Römer, F.G., Diederen, H.S.M.A., Guicherit, R., van ten Hout, K.D.: Physico-Chemical Behaviour of Atmospheric Pollutants, Proceedings of the Second European Symposium, 29 Sept.–1 Oct. 1981, Versino, B. and Ott, H. (eds.), p. 460, D. Reidel Publ. Company, Dordrecht (Netherlands), 1982
85. Muschalik, B.: Der Einfluß meteorologischer Parameter auf die bodennahe Oxidatienverteilung, in Ref. [76], Chapter VI, pp. 1–32
86. Guicherit, R., van Dop, H.: Atmos. Environ. *11*, 145 (1977)
87. Guicherit, R.: Photochemical Smog Formation in the Netherlands. TNO-Report, 1–240, TNO,s'-Gravenhage (The Netherlands), 1978
88. Apling, A.J., Sullivan, E.J., Williams, M.L., Ball, D.J., Bernard, R.E., Derwent, R.G., Eggleton, A.E.J., Hampton, L., Waller, R.E.: Nature *269*, 569 (1977)
89. Schjoldager, J.: Atmos. Environ. *13*, 1689 (1979)
90. Schjoldager, J.: J. Air Pollut. Control. Assoc. *31*, 1187 (1981)
91. Schjoldager, J.: Dovland, H., Grennfelt, P., Saltbones, J.: Photochemical Oxidants – North-western Europe 1976–79. A Pilot Project. NILU-Report Nr: 19/81, Norwegian Institute for Air Research, Lillestrom, Norway, 1981
92. Schjoldager, J.: Sivertsen, B., Hansen, J.E.: Atmos. Environ. *12*, 2461 (1978)
93. Grennfelt, P., Schjoldager, J.: Ambio *13*, 61 (1984)
94. Benarie, M., Benec'hi, A., Chuong, B.T., Menard, T.: Pollut. Atmos. *81*, 44 (1979)
95. Lopez, A., Prieur, S., Fontan, J., Kim, P.S.: Physico-Chemical Behaviour of Atmospheric Pollutants, Proceedings of the Second European Symposium, 29 Sept.–1 Oct. 1981, Versino, B. and Ott, H. (eds.), p. 362, D. Reidel Publ. Company, Dordrecht (Netherlands), 1982
96. Stangl. H., Lohse, C., Payrissat, M., Versino, B., Nicollin, B., Ottbrini, G., Rau, H.: First European Symposium on Physico-Chemical Behaviour of Atmospheric Pollutants, EEC/COST 61 bis – Meeting, Ispra, Oct. 1979, Proceedings, p. 472 (1980)
97. Zurita, E., Castro, M.: Atmos. Environ. *17*, 2213 (1983)
98. Božičević, Z., Klasinc, L., Cvitaš, T., Güsten, H.: Staub-Reinhalt. Luft *36*, 363 (1976)
99. Cvitaš, T., Güsten H., Klasinc, L.: Staub-Reinhalt. Luft *39*, 92 (1979)
100. Cvitaš, T., Klasinc, L.: Zaštita atmosphere *15*, 13 (1979)
101. Dezeljin, S., Gotovac, V., Cvitaš, T.: Kem. Ind. (Zagreb) *30*, 57 (1981)
102. Novak, I., Sabljić, A.: Kem. Ind. (Zagreb) *30*, 5 (1981)
103. Cvitaš, T., Güsten, H., Heinrich, G., Klasinc, L., Lalas, D.P., Petrakis, M.: Staub-Reinhalt. Luft *45*, 297 (1985)
104. Stasiuk, W.N., Coffey, P.E.: J. Air Pollut. Control Assoc. *24*, 564 (1974)
105. Steinberger, E.H., Ganor, E.: Atmos. Environ. *14*, 221 (1981)
106. Lonneman, W.A., Buffalini, J.J., Sella, R.L.: Environ. Sci. Technol. *10*, 374 (1976)
107. Cox, R.A., Roffey, M.J.: Environ. Sci. Technol. *11*, 900 (1977)
108. Singh, H.B., Hanst, P.L.: Geophys. Res. Letters *8*, 941 (1981)
109. Aikin, A.C., Herman, J.R., Maier, E.J., McQuillan, C.J.: Geophys. Res. *87*, 3105 (1982)
110. Singh, H.B., Salas, L.J.: Atmos. Environ. *17*, 1507 (1983)
111. Taylor, O.C.: J. Air Pollut. Control. Assoc. *19*, 347 (1969)
112. Dugger, W.M., Ting, I.P.: Phytopathology *58*, 1102 (1968)
113. Lovelock, J.E.: Ambio *6*, 131 (1977)
114. Spicer, C.W., Holdren, M.W., Keigley, G.W.: Atmos. Environ. *17*, 1055 (1983)
115. Holdren, M.W., Spicer, C.W., Hales, J.M.: Atmos. Environ. *18*, 1171 (1984)
116. Paeke, E., MacLean, M.A., Sandhu, H.S.: J. Air Pollut. Control. Assoc. *33*, 881 (1983)
117. Temple, P.J., Taylor, O.C.: Atmos. Environ. *17*, 1583 (1983)
118. Altshuller, A.P.: Measurements of the Products of Atmospheric Photochemical Reactions in Laboratory Studies and in Ambient Air-Relationship between Ozone and other Products, Atmos. Environ. *17*, 2383–2427 (1983)
119. Lonneman, W.A., Bufalini, J.J., Namie, G.R.: Environ. Sci. Technol. *16*, 665 (1982)
120. Holdren, M.W., Spicer, C.W.: Environ. Sci. Technol. *18*, 114 (1984)
121. Nieboer, H., van Ham, J.: Atmos. Environ. *10*, 115 (1976)
122. Penkett, S.A., Sandalls, F.J., Lovelock, J.E.: Atmos. Environ. *9*, 139 (1975)
123. Bos, R., Guicherit, R., Hoogeveen, A.: Sci. Total Environ. 7, 159 (1977)
124. Brasser, L.J., Guicherit, R., Huygen, C.: VDI-Berichte *270*, 25–32 (1977)

125. Penkett, S.A., Sandalls, F.J., Jones, B.M.R.: VDI-Berichte *270*, 47–54 (1977)

126. Löbel, J., Wipprecht, V., Schurath, U.: Staub-Reinhalt. Luft *40*, 243 (1980)

127. Bruckmann, P., Eynck, P.: Schriftenreihe der Landesanstalt für Immissionsschutz des Landes Nordrhein-Westfalen *49*, 19–28 (1979)

128. Nielsen, T., Samuelsson, U., Grennfelt, P., Thomsen, E.L.: Nature *293*, 553 (1981)

129. Schjoldager, J., Wathne, B.M., Brenna, D., Hov, Ø, Johannessen, T., Stige, L., Tveita, B.: "Measurements of peroxyacetyl nitrate in Oslo and southern Telemark 1980–82", Norwegian Institute for Air Research, Report 27/83, Lillestorm, Norway, 1983

130. Glavas, S., Schurath, U.: Chimika Chronika, New Series, *12*, 89 (1983)

131. Brice, K.A., Penkett, S.A., Atkins, D.H.F., Sandalls, F.J., Bamber, D.J., Tuck, A.F., Vaughan, G.: Atmos. Environ. *18*, 2691 (1984)

132. Gay, B.W., Jr., Bufalini, J.J.: Hydrogen Peroxide in the Urban Atmosphere, Advan. Chem. Ser. *113*, 255–263 (1972)

133. Bufalini, J.J., Gay, B.W., Jr., Brubaker, K.L.: Environ. Sci. Technol. *6*, 816 (1972)

134. Kok, G.L., Holler, T.P., Lopez, M.B., Nachtrieb, A.H., Yuan, M.: Environ. Sci. Technol. *12*, 1072 (1978)

135. Kok, G.L., Darnall, K.R., Winer, A.M., Pitts, J.N., Gay, B.W.: Environ. Sci. Technol. *12*, 1077 (1978)

136. Zika, R.G., Saltzman, E.S.: Geophys. Res. Letters *9*, 231 (1982)

137. Heikes. B.G., Lazarus, A.L., Kok, G.L., Kunen, S.M., Gandrub, B.W., Githin, S.N., Sperry, P.D.: J. Geophys. Res. *87*, 3045 (1982)

138. Penkett, S.A., Jones, B.M.R., Brice, K.A., Eggleton, A.E.J.: Atmos. Environ. *13*, 123 (1979)

139. Martin, L.R., Damschen, D.E.: Atmos. Environ. *15*, 1149 (1981)

140. Güsten, H.: Photocatalytic Degradation of Atmospheric Pollutants on the Surface of Metal Oxides, in: "Chemistry of Multiphase Atmospheric Systems", W. Jaeschke (ed.), Springer-Verlag, Berlin-Heidelberg-New York, 1985

141. Graham, R.A., Johnston, H.S.: J. Phys. Chem. *82*, 254 (1978)

142. Platt, U., Perner, D., Winer, A.M., Harris, G.W., Pitts, J.N., Jr.: Geophys. Res. Letters 7, 89 (1980)

143. Platt, U., Perner, D., Schröder, J., Kessler, C., Toenissen, A.: J. Geophys. Res. *86*, C 12, 11965 (1981)

144. Morris, E.D., Niki, H.J.: J. Phys. Chem. *77*, 1929 (1973)

145. Cox, R.A.: Tellus *26*, 235 (1974)

146. Cox, R.A., Coker, G.B.: J. Atmos. Chem. *1*, 53 (1983)

147. Tuazon, E.C., Winer, A.M., Pitts, J.N., Jr.: Environ. Sci. Technol. *15*, 1232 (1981)

148. Atkinson, R., Plum, C.N., Carter, W.P.L., Winer, A.M., Pitts, J.N., Jr.: J. Phys. Chem. *88*, 1210 (1984)

149. Atkinson, R., Aschmann, S.M., Winer, A.M., Pitts, J.N., Jr.: Environ. Sci Technol. *18*, 370 (1984)

150. Atkinson, R., Plum, C.N., Carter, W.P.L., Winer, A.M., Pitts, J.N., Jr.: J. Phys. Chem. *88*, 2361 (1984)

151. Winer, A.M., Atkinson, R., Pitts, J.N., Jr.: Science *224*, 156 (1984)

152. Robinson, E., Robbins, R.C.: J. Air Pollut. Control Assoc. *20*, 303 (1970)

153. Ehhalt, D.H., Drummond, J.W.: The Tropospheric Cycle of NO_x, in: "Chemistry of the Unpolluted and Polluted Atmosphere", H.W. Georgii and W. Jaeschke (eds.), pp. 219–251, D. Reidel Publ. Comp., Dordrecht – Boston – London, 1982

154. Miller, D.F., Spicer, C.W.: J. Air Pollut. Control. Assoc. *25*, 940 (1975)

155. Spicer, C.W.: Atmos. Environ. *11*, 1089 (1977)

156. Kelly, T.V., Stedman, D.H., Kok, G.L., Geophys. Res. Letters *6*, 375 (1979)

157. Kelly, T.V., Stedman, D.H., Ritter, V.A., Harvey, R.B.: J. Geophys. Res. *85*, 7417 (1980)

158. Huebert, B.J.. Geophys. Res. Letters 7, 325 (1980)

159. Klockow, D.: VDI-Berichte *429*, 165 (1982)

160. Slanina, J.: VDI-Berichte *429*, 177 (1982)

161. Perner, D., Platt, U.: Geophys. Res. Letters *6*, 917 (1979)

162. Platt, U., Perner, D.: J. Geophys. Res. *85*, 7435 (1980)

163. Harris, G.W., Carter, W.P.L., Winer, A.M., Pitts, J.N., Jr.: Environ. Sci. Technol. *16*, 414 (1982)
164. Nash, T.: Tellus *26*, 175 (1974)
165. Chan, W.H., Nordstrom, R.J., Calvert, J.G., Shaw, J.H.: Environ. Sci. Technol. *10*, 674 (1976)
166. Kaiser, E.W., Wu, C.H.: J. Phys. Chem. *81*, 1701 (1977)
167. Platt, U.: The Origin of Nitrous and Nitric Acid in the Atmosphere, in: "Chemistry of Multiphase Atmospheric System", W. Jaeschke (ed.), Springer-Verlag, Berlin-Heidelberg-New York, 1985
168. Pitts, J.N., Jr., Winer, A.M., Atkinson, R., Carter, W.P.L.: Environ. Sci. Technol. *17*, 54 (1983)
169. Pitts, J.N., Jr., Grosjean, D., van Cauwenberghe, K., Schmid, J.P., Fritz, D.R.: Environ. Sci. Technol. *12*, 946 (1978)
170. National Research Council: Formaldehyde and other aldehydes. Committee on Aldehydes, Board on Toxicology and Environmental Health Hazards, National Academy Press, Washington, D.C., 1981
171. Platt, U., Perner, D., Patz, H.W.: J. Geophys. Res. *84*, 6329 (1979)
172. Neitzert, V., Seiler, W.: Geophys. Res. Letters *8*, 79 (1981)
173. Kopczynski, S.L., Altshuller, A.P., Sutterfield, F.D.: Environ. Sci. Technol. *8*, 909 (1974)
174. Grosjean, D., Wright, B.: Atmos. Environ. *17*, 2093 (1983)
175. Doyle, G.J.: Environ. Sci. Technol. *4*, 907 (1970)
176. Winer, A.M., Graham, R.A., Doyle, G.J., Bekowies, P.J., McAfee, J.M., Pitts, J.N., Jr.: Advan. Environ. Sci. Technol. *10*, 461 (1980)
177. Niki, H., Maker, P.D., Savage, C.M., Breitenbach, L.P.: in "Studies in physical and theoretical chemistry, Vol. 8, pp. 1–13, F.J. Comes, A. Müller and W.J. Orville-Thomas (eds.), Elsevier Scientific Publ. Comp., Amsterdam – Oxford – New York, 1980
178. Hull, L.A.: in "Proc. Symp. on Atmospheric Biogenic Hydrocarbons: Emission Rates Concentrations and Fates (J.J. Bufalini and R.R. Arnts, eds.), pp. 482–518, Research Triangle Park, NC., 8–9 Jan. 1980
179. Spicer, C.W., Jones, P.W.: J. Air Pollut. Control Assoc. *11*, 1122 (1977)
180. Tagaki, H., Washida, N., Akimota, H., Nagasawa, K., Usul, Y., Okura, M.: J. Phys. Chem. *84*, 478 (1980)
181. Hanst, P.L., Wong, N.W., Bragin, J.: Atmos. Environ. *16*, 969 (1982)
182. Su, F., Calvert, J.G., Shaw, J.H.: J. Phys. Chem. *83*, 3185 (1979)
183. Van Duuren, B.L.: Intern. J. Environ. Anal. Chem. *1*, 233 (1972)
184. Hanst, P.L., Gay, B.W., Jr.: Atmos. Environ. *17*, 2259 (1983)
185. Rubino, R.A., Bruckman, L., Magyar, J.: J. Air Pollut. Control Assoc. *26*, 972 (1976)
186. Lalas, D.P., Asimakopoulos, D.N., Deligiorgi, D.G., Helmis, C.G.: Atmos. Environ. *17*, 1621 (1983)
187. Aldaz, L.: J. Geophys. Res. *74 C*, 6943 (1969)
188. Garland, J.A., Elzerman, A.W., Penkett, S.A.: J. Geophys. Res. *85 C*, 7488 (1980)
189. Cvitaš, T., Güsten, H., Heinrich, G., Klasinc, L., Lalas, D.P., Petrakis, M.: (to be published)
190. Lyons, W.A., Cole, H.S.: J. Appl. Met. *15*, 733 (1976)
191. Hov, Ø., Hesstvedt, E., Isaksen, I.S.A.: Nature *273*, 341 (1978)
192. Karl, T.R.: Atmos. Environ. *12*, 1421 (1978)
193. Senum, G.I., Lee, Y.N., Gaffney, J.S.: J. Phys. Chem. *88*, 1269 (1984)
194. Wallington, T.J., Atkinson, R., Winer, A.M.: Geophys. Res. Letters *11*, 861 (1984)
195. Elkus, B., Wilson, K.R.: Atmos. Environ. *11*, 509 (1977)
196. Cleveland, W.S., Graedel, T.E., Kleiner, B., Warner, J.L.: Science *186*, 1037 (1974)
197. Wolff, G.T., Lioy, P.J., Wright, G.D., Meyers, R.E., Cederwall, R.T.: Atmos. Environ. *11*, 797 (1977)
198. Spicer, C.W., Joseph, D.W., Sticksel, P.R., Ward, G.F.: Environ. Sci. Technol. *13*, 975 (1979)
199. Pratt, G.C., Hendrickson, R.C., Chevone, B.I., Christopherson, D.A., O'Brien, M.V., Krupa, S.V.: Atmos. Environ. *17*, 2013 (1983)

200. Worth, J.J.B., Ripperton, L.A.: Rural Ozone – Sources and Transport, Advan. Environ. Sci. Engng. *3*, 150–170 (1980)
201. Clarke, J.F., Ching, J.K.S.: Atmos. Environ. *17*, 1703 (1983)
202. Reynolds, S.D., Roth, P.M., Seinfeld, J.H.: Atmos. Environ. *7*, 1033 (1973)
203. Lamb, R.G., Seinfeld, J.H.: Environ. Sci. Technol. *7*, 253 (1973)
204. Pasquill, F.: "Atmospheric Diffusion", Van Nostrand, London, 1962
205. Hanna, S.: Atmos. Environ. *7*, 803 (1973)
206. Venkatram, A.: Atmos. Environ. *12*, 2243 (1978)
207. Shreffler, J.H., Schere, K.L.: Evaluation of four urban-scale photochemical air quality simulation models. EPA-Report 600/3-82-043 (1982)
208. Scherer, B., Stern, R.: Physico-chemical Behaviour of Atmospheric Pollutants, Proceedings of the Second European Symposium, 29 Sept.–1 Oct. 1981, Versino, B. and Ott, H. (eds.), p. 561, D. Reidel Publ. Company, Dordrecht (Netherlands), 1982
209. Graedel, T.E., Farrow, L.A., Weber, T.A.: Atmos. Environ. *10*, 1095 (1976)
210. Reynolds, S.D., Ames, J., Hecht, T.A., Meyer, J.P., Whitney, D.C.: Continued Research in Mesoscale Air Pollution Simulation Modeling. Vol. II. EPA-Report 600/4-76-016 d (1976)
211. McCracken, M., Wuebbles, D., Walton, J., Duewer, W., Grant, K.: J. Appl. Meteorol. *17*, 254 (1978)
212. Schiavone, J.A., Graedel, T.E.: Atmos. Environ. *15*, 163 (1981)
213. Liu, M.K., Seinfeld, J.H.: Atmos. Environ. *9*, 553 (1975)
214. McRae, G.J., Goodin, W.R., Seinfeld, J.H.: Mathematical Modelling of Photochemical Air Pollution, EQL-Report No. 18, pp. 1–661 (1982)
215. Atkinson, R., Lloyd, A.C.: Evaluation of Kinetic and Mechanistic Data for Modelling of Photochemical Smog, J. Phys. Chem. Ref. Data *13*, 315–444 (1984)
216. Proceedings of the EPA-OECD International Conference on Long-Range Transport Models for Photochemical Oxidants and Their Precursors, EPA-Report 600/9-84-006, Environmental Sciences Research Laboratory, U.S. Environmental Protection Agency, Research Triangle Park, North Carolina, 1984
217. Air Quality Criteria for Ozone and Other Photochemical Oxidants, EPA-Report 600/8-78-004. United States Environmental Protection Agency, Research Triangle Park, N.C., 1978
218. Von Nieding, G., Wagner, H.M.: Wirkungen von Photooxidantien auf Mensch und Tier, in: Luftqualitätskriterien für photochemische Oxidantien, Report 5/83, pp. 133–203, Umweltbundesamt, E. Schmidt Verlag, Berlin, 1983
219. Gilette, D.G.: J. Air Pollut. Control Assoc. *27*, 329 (1977)
220. Heck, W.W., Taylor, O.C., Adams, R., Bingham, G., Miller, J., Preston, E., Weinstein, L.: J. Air Pollut. Control Assoc. *32*, 353 (1982)
221. Jacobson, J.S.: VDI-Berichte *270*, 163 (1977)
222. Guderian, R., Tingey, D.T., Rabe, R.: Wirkungen von Photooxidantien auf Pflanzen, in: Luftqualitätskriterien für photochemische Oxidantien, Report 5/83, pp. 205–427, Umweltbundesamt, E. Schmidt Verlag, Berlin 1983
223. Haynie, F.H., Spence, J.W.: J. Air Pollut. Control Assoc. *34*, 941 (1984)
224. Shaver, C.L., Cass, G.R., Druzik, J.R.: Environ. Sci. Technol. *17*, 748 (1983)
225. Schreiber, H.: Wirkungen von Photooxidantien auf Materialien, in: Luftqualitätskriterien für photochemische Oxidantien, Report 5/83, pp. 429–443, Umweltbundesamt, E. Schmidt Verlag, Berlin, 1983
226. Jessel, U.: VDI-Berichte *270*, 145 (1977)
227. Takeuchi, K., Yazawa, T., Ibusuki, T.: Atmos. Environ. *17*, 2253 (1983)
228. Dlugi, R., Güsten, H.: Atmos. Environ. *17*, 1765 (1983)
229. Ludwig, F.L., Javitz, H.S., Valdés, A.: J. Air Pollut. Control Assoc. *33*, 963 (1983)
230. Dodge, M.C.: Combined Use of Modeling Techniques and Smog Chamber Data to Derive Ozone-Precursor Relationships, Proceedings of the Intern. Conf. on Photochemical Oxidant Pollution and its Control, EPA-Report 600/3-77-001b, pp. 881–889 (1979)
231. Kinosian, J.R.: Environ. Sci. Technol. *16*, 880 (1982)
232. Jones, K.H., Ruch, R.B., Barone, J.B., Walsh, J.F., Karpovich, R.A.: J. Air Pollut. Control Assoc. *33*, 330 (1983)

233. Laird, A.R., Miksad, R.W., Middleton, P.: J. Air Pollut. Control Assoc. *32*, 1221 (1982)
234. Simpson, R.W., Layton, A.P.: Atmos. Environ, *17*, 1649 (1983)
235. Stephens, E.R., Linnell, R.H., Reckner, L.: Science *138*, 831 (1962)
236. Maugh II, T.H.: Science *193*, 871 (1976)
237. Tinker, J.: New Scientist *72*, 530 (1976)
238. Jayanti, R.K.M., Simonaitis, R., Heicklen, J.: Atmos. Environ. *8*, 1283 (1974)
239. Gorse Jr., R.A., Lii, R.R., Saunders, B.B.: Science *197*, 1365 (1977)
240. Filby, W.G., Güsten, H.: Atmos. Environ. *12*, 1563 (1978)
241. Filby, W.G., Güsten, H.: Atmos. Environ. *13*, 424 (1979)
242. Münzer R., Filby, W.G.: Chemosphere *10*, 809 (1979)
243. Penzhorn, R.D., Filby, W.G., Güsten, H.: Z. Naturforsch. *29a*, 1449 (1974)
244. Ashmore, M.R., Bell, J.N.B., Reily, C.L.: Nature *276*, 813 (1978)
245. Ozone and other Photochemical Oxidants, National Academy of Sciences, National Research Council, Washington, D.C. (1976)
246. Karl, T.R.: Atmos. Environ. *14*, 681 (1980)
247. Miller, P.R., McCutcheon, M.H., Milligan, H.D.: Atmos. Environ. *6*, 623 (1972)
248. Varfalvy, L., Jegier, Z.: Chemosphere *3*, 35 (1974)
249. Quickert, N., Dubois, L.: Sci. Total Environ. *2*, 81 (1973)
250. Bottenheim, J.W., Braslavsky, S.E., Strausz, O.P.: Environ. Sci. Technol. *11*, 801 (1977)
251. Bravo, H., Magaña, A., Magaña, R.: Staub Reinhalt. Luft *39*, 427 (1979)
252. Wakamatsu, S., Ogawa, Y., Murano, K., Goi, K., Abutamoto, Y.: Atmos. Environ. *17*, 827 (1983)
253. Galbally, I.E.: Atmos. Environ. *5*, 15 (1971)
254. Post, K., Bilger, R.W.: Atmos. Environ. *12*, 1857 (1978)
255. Post, K.: Atmos. Environ. *13*, 783 (1979)
256. DOE-Report NBB-0044, March 1983, National Tech. Inform. Service, U.S. Departm. of Commerce, Springfield, Virginia

Atmospheric Distribution of Pollutants
and Modelling of Air Pollution Dispersion

H. van Dop

Royal Netherlands Meteorological Institute
P.O. Box 201, NL-3730 AE De Bilt, The Netherlands

Summary

This chapter discusses the major mechanisms which cause the dispersion of pollution in the atmosphere. First a brief review is given of elementary atmospheric physics and hydrodynamics. The major source of energy for atmospheric motion is the solar radiation. It is converted into atmospheric kinetic energy (mean circulation and turbulence) by a chain of processes. Since the dispersion of pollution is mostly limited to the lowest atmospheric layer (the atmospheric boundary layer) the attention is focussed on the physical structure and dynamics in this layer. Then follows the introduction of the dispersion equation, which is essentially a conservation equation. Also some alternative approaches are formulated and methods are indicated to obtain their analytical or numerical solution. Finally a summary of air pollution models is given, where a distinction is made between urban models on a scale less than 50 km, medium range models up to a scale of ~ 500 km and long range transport models, which operate on the continental scale.

Introduction

Inherent to all kinds of processes is the production of atmospheric admixtures. They are related to various activities of our modern, industrial society, but also natural processes may contribute considerably. In densely populated and highly industrialised areas the production of waste may lead to undesired effects on human health, animal life and vegetation, or to enhanced corrosion of materials and goods. The reduction or elimination of these effects is a matter of policy, which, in order to be effective, requires the knowledge of causal relationships between emissions of waste on the one hand and their ultimate effects on the environment on the other.

In the case where waste products are released in the atmosphere in the form of gaseous or particulate matter these relationships are far from trivial because of the highly unpredictable ways pollutants travel. The atmosphere is in constant, turbulent motion, and transport of airbone material can only be determined in a rough statistical way. The problem of transport of air pollution in the atmosphere is thus closely related to the atmospheric motion itself, since we consider mainly pollutants which do not influence the atmospheric dynamics (passive pollutants).

To determine the atmospheric distribution of pollutants requires a thorough knowledge of (the statistics of) the atmospheric motion which, by its turbulent nature, involves a broad range of time- and spatial scales. Dispersion of pollution over a distance of a few kilometres for example, takes place under conditions which may be much different from dispersion considered on a continental or global scale. It is therefore natural that in the mathematical description of air pollution dispersion one distinguishes between various ranges, and we will do so here. First however, some topics in atmospheric physics and dynamics will be introduced, which are essential for the understanding of atmospheric modelling.

Atmospheric Physics

Atmospheric Radiation*

The primary energy source of the atmosphere is the solar radiation. The radiation energy flux (entering the atmosphere through an imaginary plane perpendicular to

---

*   A more detailed treatment of this subject is given by H. J. Bolle in Vol. IB, p. 131 of the handbook

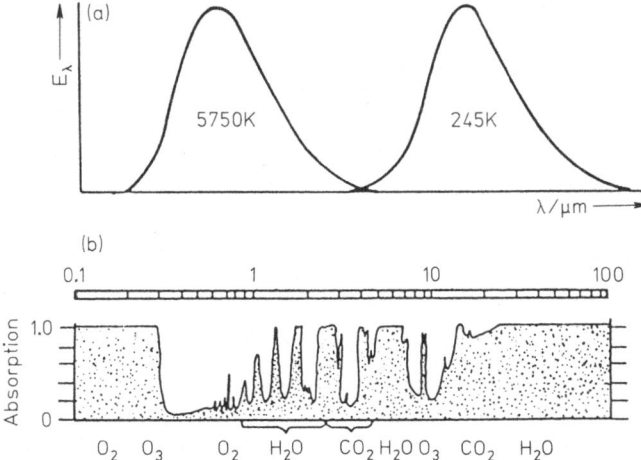

Fig. 1. (a) The spectral energy distribution of the radiation of the sun (5750 K) and atmosphere (245 K). The areas under the curves (the total radiated energy, Eq. (2), are approximately equal suggesting an approximate balance. (b) The atmospheric absorption is also indicated as a function of wavelength. Note that the atmosphere (no clouds) is almost transparent in the wavelength region where solar radiation has its maximum intensity. It also shows that the atmosphere is almost opaque in the region where the earth emits the major part of its radiation. (Source: J. T. Houghton, The physics of atmospheres, Cambridge Univ. Press 1977)

the incoming direction), amounts to $1400 \ \mathrm{Jm}^{-2} \mathrm{s}^{-1}$, the solar constant (S). Most of the radiation energy is emitted in the visible range ($\lambda = 0.4 – 0.7 \ \mu m$).

The solar spectrum is fairly accurately given by Planck's law, which relates the energy flux at a given wavelength (E_λ) to the temperature (T) of a black body according to (Fig. 1)

$$E_\lambda = \frac{2\pi c^2 h \lambda^{-5}}{\exp(ch/\lambda kT) - 1},$$

(1)

where c is the speed of light ($3 \ 10^8 \ \mathrm{ms}^{-1}$), h, Planck's constant ($6.63 \ 10^{-34} \ \mathrm{Js}$) and k Boltzmann's constant ($1.4 \ 10^{-23} \ \mathrm{JK}^{-1}$). Approximately 30% of the solar radiation is reflected into space (the Albedo, A), the other part is absorbed by the atmosphere and earth (Fig. 2). They also emit radiation according to the same law (1), but due to the much lower temperatures of the earth and atmosphere, at a much lower wavelength (5–25 µm), the "far infrared". When it is assumed that there is a balance in radiation, one may obtain a crude estimate of the earth-atmosphere temperature by equating the incoming short wave radiation and the outgoing long wave radiation. The latter is found by integrating (1) over all wave lengths, yielding

$$E_T = \sigma T^4,$$

(2)

(Wien's law), where σ is $5.67 \ 10^{-8} \ \mathrm{Jm}^{-2} \mathrm{s}^{-1} \mathrm{K}^{-4}$. Relating this to the incoming radiation we obtain by a simple geometric consideration

$$(1 - A)S = 4\sigma T^4,$$

(3)

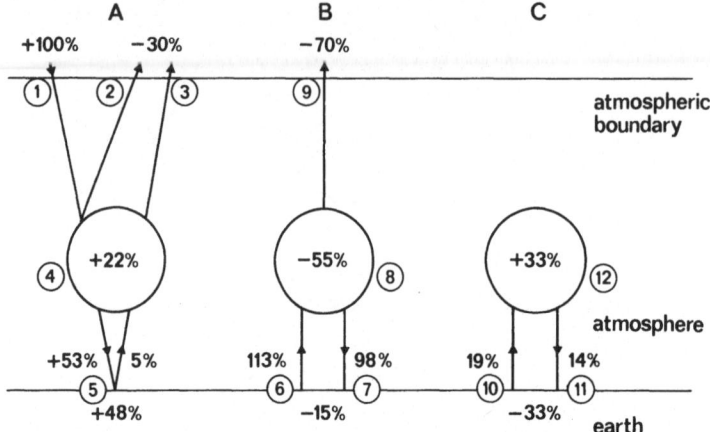

Fig. 2. Average yearly radiation and energy budget of the earth and atmosphere. (A) solar radiation (short wavelength), (B) long wave (terrestrial and atmospheric) radiation, (C) energy budget. 1. incoming solar radiation, normalised at 100%. 2,3. reflection by clouds and earth surface (Albedo). 4. total atmospheric absorption. 5. terrestrial absorption. 6. emitted terrestrial radiation. 7. absorbed atmospheric radiation. 8. total emitted atmospheric radiation. 9. total radiation loss of earth-atmosphere. 10. latent heat flux into the atmosphere. 11. sensible heat flux into the atmosphere. 12. total absorbed heat by atmosphere. The energy balance for the atmosphere for example is as follows: solar radiation (22%) and sensible and latent heat (33%) balance the (net) loss of long wave radiation (55%). (Source: luchtverontreiniging en weer, Staatsuitgeverij, Den Haag 1979)

from which follows $T = 256$ K. On the average there is no perfect balance at the earth's surface between the incoming short wave and outgoing long and short wave radiation. There is a surplus of radiation ($\sim 30\%$), which is converted into heat. It is carried away from the surface by three convective processes: the soil heat flux (G), the sensible heat flux (H) and the water vapour flux (E). Assuming an approximate balance between these quantities we obtain

$$Q^* = G + H + \lambda E, \tag{4}$$

where Q^* is the net radiation flux and λ the latent heat of evaporisation. The soil heat flux is usually small during daytime ($< 10\%$). The ratio of sensible over latent heat flux is often referred to as the Bowen ratio, $B \equiv H/\lambda E$.

The atmospheric fluxes H and E are important sources of energy in the lower atmosphere. Also they largely determine the temperature in this atmospheric layer which is, as we shall see, strongly coupled with the atmospheric dynamics. First, however, some concepts in atmospheric thermodynamics will be discussed.

Thermodynamics

Because of the presence of the earths' gravitational field the density of the atmosphere decreases with altitude. When vertical accelerations are neglected, it follows from the static equilibrium condition that the fall in pressure, dp, over a small increment in height (dz) is given by

$$dp = -\varrho g\, dz, \qquad (5)$$

where $\varrho = \varrho(z)$ is the density of air and g the gravitational acceleration. We consider a parcel of air and neglect radiation processes. The first law of thermodynamics for a unit mass is

$$dQ = c_v dT + pd\left(\frac{1}{\varrho}\right), \qquad (6)$$

where c_v is the specific constant of heat at constant volume. When we neglect heat exchange of the parcel with surrounding air, $dQ = 0$ (the adiabatic condition). Combining (5 and 6) with the equation of state,

$$p = \varrho\, \frac{R}{m_a}\, T, \qquad (7)$$

where R is the universal gas constant $(8.31\,\mathrm{JK^{-1}\,Mol^{-1}})$ and m_a the (mean) molecular weight of air, we obtain

$$\frac{dT}{dz} = -\frac{g}{c_p} \equiv -\Gamma_d, \qquad (8)$$

where Γ_d is the dry adiabatic lapse rate. The specific heat at constant pressure, c_p is equal to $c_v + R/m_a$ $(=1005\,\mathrm{JKg^{-1}\,K^{-1}})$. Equation (8) indicates that in a dry adiabatic atmosphere the temperature drops with approximately $10^{-2}\,\mathrm{K/m}$. In an atmosphere where moderate to fast (more or less random) motions occur, so that they can be considered adiabatic, the temperature gradient or lapse rate will be given by (8). However, in an atmospheric layer, which by some cause has a lapse rate which is different from (8), vertical motion will be subject to buoyancy forces, due to temperature (density) differences between rising (or falling) parcels of air and their surrounding. This is illustrated in Fig. 3, where a parcel of air is considered in an environment where $dT/dz > -\Gamma_d$ (Fig. 3a) and $dT/dz < -\Gamma_d$ (Fig. 3b) respectively. In the former case a parcel which moves upwards adiabatically, enters a warmer (less dense) region, and will be forced downwards again to its initial position. By the same reasoning a restoring force will act on it when moving downwards. In the other case (Fig. 3b) parcels moving upwards or downwards undergo an acceleration, which reinforces the motion, so that they tend to move away from their initial positions, a situation which is referred to as *unstable*. The first case is referred to as *stable*. An important conclusion is that obviously the vertical temperature distribution influences the atmospheric dynamics: in a stable atmosphere vertical motions are damped, whereas in an unstable atmosphere vertical motions are enhanced. This of course has large consequences for the (vertical) dispersion of pollution, when moving air parcels carry a load of contaminant material.

Near the surface, the surface temperature has a large impact on the atmospheric stability: during nighttime the surface cools by radiative processes, and a shallow stable layer develops; during daytime direct and indirect solar radiation heats the surface, which induces an unstable atmospheric boundary layer. In view of its important consequences for atmospheric dispersion, a measure

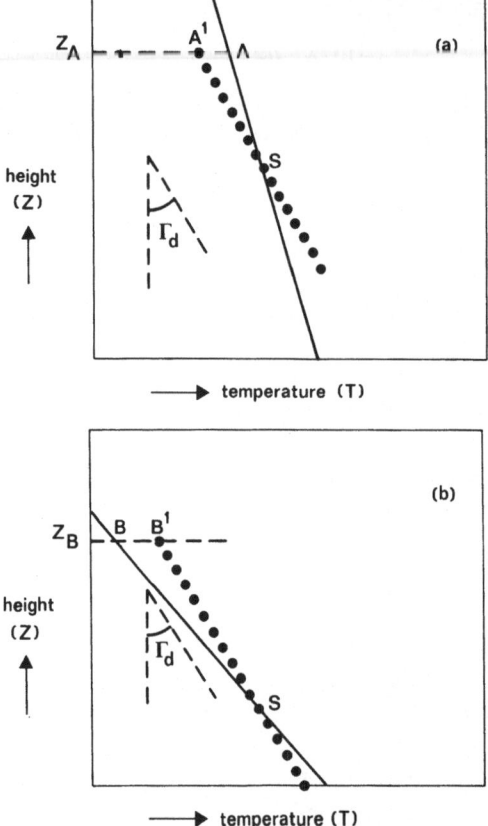

Fig. 3. A schematic explanation of atmospheric stability. In the figure are indicated two actual atmospheric temperature profiles (solid lines) and the temperature of an adiabatically moving parcel of air (dotted line), which slopes, according with (8), as Γ_d. S denotes the initial position of the air parcel, from which it is moved up- or downwards. When $dT/dz > -\Gamma_d$ (a) we observe that at Z_A the temperature of the air parcel ($T_{A'}$) is lower than that of the surrounding air (T_A) and thus $\varrho_{A'} > \varrho_A$ (because $p_A = p_{A'}$). Hence the gravitational acceleration forces the particle back to its initial position, introducing a (damped) harmonic motion along the dotted curve. When $dT/dz < -\Gamma_d$ (b) we observe that $T_{B'} > T_B$ at Z_b, so that buoyancy now reinforces up- (or downward) motion.

for the atmospheric stability is desired and if possible should be estimated from routine meteorological measurements (rather than from not routinely available radiation data). A convenient scheme for atmospheric stability was derived by Pasquill [1], who distinguished six stability categories (A to F), see Table 1. These categories play an important role in the vertical and horizontal dispersion of pollutants released near the surface, as we shall see in the next sections.

The assumption that vertical atmospheric motions may be considered adiabatic leads to the concept of potential temperature. From (6) and (7) we have

$$dS \equiv \frac{dQ}{T} = c_p \frac{dT}{T} - \frac{R}{m_a} \frac{dp}{p}, \qquad (9)$$

Table 1. The determination of stability class over land from routine weather data (KNMI). (Wind speed V in knots (1 knot = 0.5 m/s), cloud cover, N in octa's). The day is defined as the period between one hour after sunrise and one hour before sunset.

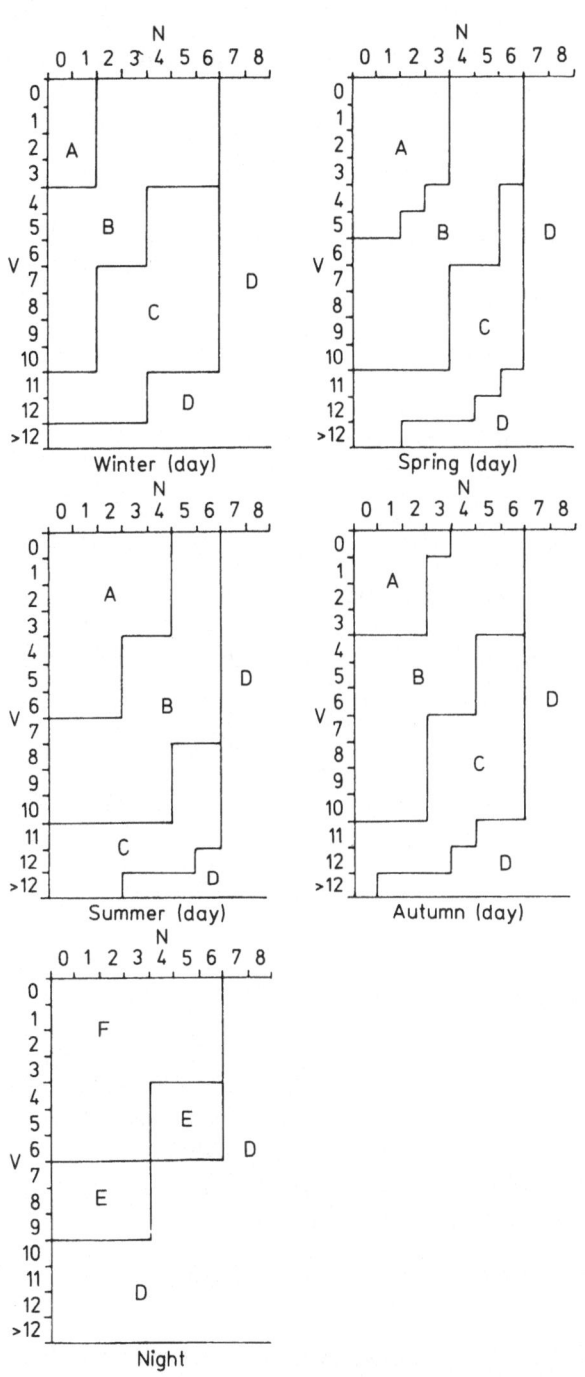

where S denotes the entropy. This can be integrated to

$$S = c_p \ln T - \frac{R}{m_a} \ln p + C, \tag{10}$$

where C is an integration constant. Noting that for adiabatic motions S = constant we obtain from (1)

$$c_p \ln \frac{T_2}{T_1} = \frac{R}{m_a} \ln \frac{p_2}{p_1}, \tag{11}$$

where the indices 1, 2 refer to any two levels in the atmosphere. We now define the potential temperature, θ, of a parcel of air as its temperature at the standard pressure level of 1000 mbar (p_0) or,

$$c_p \ln \frac{T}{\theta} = \frac{R}{m_a} \ln \frac{p}{p_0}, \tag{12}$$

which, using $c_p - c_v = R/m_a$, can be written as

$$\theta = T \left(\frac{p_0}{p} \right)^\kappa, \tag{13}$$

where $\kappa = (c_p - c_v)/c_p$. Note that apart from a constant, θ is equivalent with the entropy S.

Generally by radiation and horizontal advection the vertical temperature distribution of the atmosphere will be non-adiabatic, so that from (13)

$$\frac{d \ln \theta}{dz} = \frac{d \ln T}{dz} - \kappa \frac{d \ln p}{dz}, \tag{14}$$

which can be rewritten as

$$\frac{d\theta}{dz} = \frac{\theta}{T} \left(\frac{dT}{dz} + \Gamma \right) \simeq \frac{dT}{dz} + \Gamma_d. \tag{15}$$

For most applications in air pollution dispersion only the lowest 3 km of the atmosphere, where $\theta \simeq T$, is relevant, so that the approximate relationship holds quite accurately.

The same considerations hold in an atmosphere which contains water vapour and liquid water. Parcels moving upwards go down in temperature according to (8), until condensation occurs. The condensation heat is transferred to the parcel so that its temperature decrease is less than it would be in the case of dry air. This process continues until the total water vapour content of the parcel is condensated. The adiabatic lapse rate, Γ_w, is given by

$$\Gamma_w = \Gamma_d (1 + \alpha) \left(1 + \alpha \frac{\varepsilon \lambda}{c_p T} \right)^{-1}, \tag{16}$$

where $\alpha = \lambda E m_a / pRT$ and $\varepsilon = m_w/m_a$, the ratio of molecular weight of air and water vapour.

Atmospheric Dynamics

The major part of the air pollution is released at or near the earth's surface. It is transported by the large scale, mainly horizontal motion and diffuses gradually to more elevated layers by smaller scale, more or less random motions. Most air pollution, however, is confined within a layer with a thickness of less than three km. The atmospheric dynamics and turbulence within this layer are strongly influenced by the surface of the earth. They are described by the Navier-Stokes equations. We refer to textbooks on geophysical fluid dynamics [2–4] for a full treatment of this subject. Here we shall only present some useful results within the context of the dispersion of air pollution.

The Fluid Dynamics Equations

The circulation in the atmospheric boundary layer is driven by pressure differences. Further, the presence of the surface makes that the wind velocity drops to zero near the surface. Another force which acts on an air parcel is friction, while one also has to take into account that the air flow is usually considered in a frame of reference which is fixed to the earth, so that an apparent rotational force acts on the flow (the Coriolis force). The equations of motion in a suitable approximation are

$$\frac{du}{dt} = -\frac{1}{\varrho}\frac{\partial p}{\partial x} + fv + v\frac{\partial^2 u}{\partial z^2},$$

$$\frac{dv}{dt} = -\frac{1}{\varrho}\frac{\partial p}{\partial y} - fu + v\frac{\partial^2 v}{\partial z^2}, \tag{17}$$

$$\frac{dw}{dt} = -\frac{1}{\varrho}\frac{\partial p}{\partial z} - g + v\frac{\partial^2 x}{\partial z^2},$$

where d/dt denotes $\partial/\partial t + u\partial/\partial x + v\partial/\partial y + w\partial/\partial z$. The viscosity is given by v, the pressure by p, the Coriolis parameter by f, and the components of the wind velocity in easterly, northerly, and upward direction by u, v, and w respectively. Together with the continuity equation, which is in a good approximation given by

$$\frac{\partial u}{\partial x} + \frac{\partial v}{\partial y} + \frac{\partial w}{\partial z} = 0 \tag{18}$$

and the thermodynamic equations (6) and (7), they determine the fluid flow. Usually Eqs. (17) are solved by decomposing all dynamic variables into a mean and a fluctuating component, for example $u = U + u'$, etc., whence,

$$\frac{DU}{Dt} = -\frac{1}{\varrho}\frac{\partial P}{\partial x} + fV - \left(\frac{\overline{\partial u'^2}}{\partial x} + \frac{\overline{\partial u'v'}}{\partial y} + \frac{\overline{\partial u'w'}}{\partial z}\right),$$

$$\frac{DV}{Dt} = -\frac{1}{\varrho}\frac{\partial P}{\partial y} - fU - \left(\frac{\overline{\partial v'u'}}{\partial x} + \frac{\overline{\partial v'^2}}{\partial y} + \frac{\overline{\partial v'w'}}{\partial z}\right), \tag{19}$$

$$\frac{DW}{Dt} = -\frac{1}{\varrho}\frac{\partial P}{\partial z} - g - \left(\frac{\overline{\partial w'u'}}{\partial x} + \frac{\overline{\partial w'v'}}{\partial y} + \frac{\overline{\partial w'^2}}{\partial z}\right),$$

where $D/Dt = \partial/\partial t + U\partial/\partial x + V\partial/\partial y + W\partial/\partial z$. The terms involving viscosity are neglected here. The overbars denote a time average over typically one hour. The terms in parentheses in (19) are the derivatives of the Reynold stresses and represent the flux of momentum due to the fluctuating motions.

The Ekman Layer

A particularly illustrative solution to (19) can be obtained by assuming stationarity, horizontal homogeneity of the windfield and by putting $\overline{u'w'} = -K\partial U/\partial z$ and $\overline{v'w'} = -K\partial V/\partial z$. The eddy viscosity K is a turbulence parameter and taken constant here. Neglecting vertical motion we obtain from (19)

$$0 = -\frac{1}{\varrho}\frac{\partial P}{\partial x} + fV + K\frac{\partial^2 U}{\partial z^2},$$

$$0 = -\frac{1}{\varrho}\frac{\partial P}{\partial y} - fU + K\frac{\partial^2 V}{\partial z^2}, \tag{20}$$

$$0 = -\frac{1}{\varrho}\frac{\partial P}{\partial z} - g.$$

the Ekman layer equations. The geostrophic wind field approximation (U_g, V_g) is obtained by neglecting the stresses in (20), or

$$U_g \equiv -\frac{1}{\varrho f}\frac{\partial P}{\partial y} \quad \text{and} \quad V_g \equiv \frac{1}{\varrho f}\frac{\partial P}{\partial x}. \tag{21}$$

Equations (20) can be easily solved with the boundary conditions $U = V = 0$ for $z = 0$, and $V_g = 0$ at the top of the layer. The solution is (Fig. 4)

$$U = U_g\{1 - e^{-z/h}\cos(z/h)\},$$

$$V = U_g e^{-z/h}\sin(z/h), \tag{22}$$

where

$$h = (2K/f)^{1/2}. \tag{23}$$

Above the level $z = \pi h$ the wind is approximately geostrophic. Therefore this level may be considered as the atmospheric boundary layer height, which for $f \sim 10^{-4}\,\mathrm{s}^{-1}$ and $K = 10\,\mathrm{m}^2\,\mathrm{s}^{-1}$, is approximately 1400 m. The solution (22) also shows that apart from the increase in speed the wind direction changes (clockwise) with height, a phenomenon which is often observed, also in more complicated flows, but generally over smaller angles than the by (22) predicted $45°$.

Turbulence

In the Ekman solution the atmospheric turbulence was represented by the (constant) eddy viscosity, K. This is only a very crude approximation and in reality K depends on the turbulence in a more complicated way. Turbulence can be generated by wind shear in the vicinity of a surface, but also by motions induced by density differences (buoyancy). They are due to local heating or cooling of air

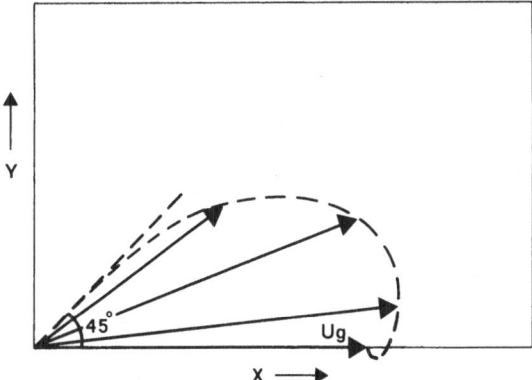

Fig. 4. "The Ekman spiral" (dashed curve). Also shown are some wind velocity vectors at four heights indicating that the wind speed increases with height and its directions turns clockwise towards the geostrophic wind direction.

parcels in the earth's gravitational field and they may generate or suppress turbulent kinetic energy. Radiative and condensation processes may affect the temperature of moist air parcels, but the dominant heating or cooling processes in the boundary layer occur by energy exchange processes at the surface. The above considerations already indicate that the turbulent kinetic energy depends on the roughness, heatflux, and altitude. This is illustrated by the equation for the mean turbulent kinetic energy $\bar{e} \equiv \frac{1}{2}(\overline{u'^2} + \overline{v'^2} + \overline{w'^2})$, which can be derived from (17):

$$\frac{\partial \bar{e}}{\partial t} = -\overline{u'w'} \frac{\partial U}{\partial z} + \frac{g}{T} \overline{w'\theta'} - \frac{\partial}{\partial z}\left(\overline{w'e'} + \frac{\overline{w'p'}}{\varrho_0}\right) - \varepsilon, \tag{24}$$

where ε is the dissipation rate of turbulent kinetic energy (tke) due to viscous processes. We have for simplicity considered the case of horizontal homogeneity. The third term on the right represents the vertical transport of tke, the second term the production (or loss) due to buoyancy and the first term is the production of tke due to wind shear. Note that the Coriolis force, only present because we describe the motion in a non-inertial frame does not contribute to the production of tke.

A further simplification is obtained by assuming stationarity ($\partial \bar{e}/\partial t = 0$) and neglecting the third term at the right of (24). First we consider the case without buoyancy ($\overline{w'\theta'} = 0$). Then we have simply

$$\varepsilon = -\overline{u'w'} \frac{\partial U}{\partial z}, \tag{25}$$

which says that the turbulence produced by the wind shear is balanced by the viscous dissipation, ε. The latter quantity can be estimated to be equal to u_*^3/kz near the surface, where u_*, the friction velocity is defined as

$$u_*^2 = \lim_{z \to 0} -\overline{u'w'} \tag{26}$$

Table 2. Roughness length for some terrain categories

Terrain category	z_0(m)
Water surface	0.0005
Open field	0.03
Field with scattered	0.25
Trees and hedges	
Roads/railways	0.50
Forest	1.00
Built-up areas	2.00

and k is Von Karman's constant ($\simeq 0.35$). Thus near the surface we have from (25)

$$\frac{\partial U}{\partial z} = \frac{u_*}{kz} \tag{27}$$

or

$$U = \frac{u_*}{k} \ln\left(\frac{z - z_0}{z_0}\right). \tag{28}$$

The constant of integration, z_0, the roughness length, depends on the roughness of the surface and can be determined experimentally. In Table 2 we give some typical values of z_0 for some terrain types.

Equation (28) indicates that close to the surface and in neutral conditions the wind profile is logarithmic. The range of validity of (28) is uncertain, but generally it does not extend beyond $z \simeq 100$ m.

When buoyancy plays a role ($\overline{w'\theta'} \neq 0$) we have to consider the approximate equality (cf. (24))

$$-\overline{u'w'}\frac{\partial U}{\partial z} + \frac{g}{T}\overline{w'\theta'} = \varepsilon, \tag{29}$$

which indicates that the shear- and buoyancy production is balanced by the viscous dissipation. Note that $\overline{u'w'}\partial U/\partial z$ is always positive whereas $g/T\overline{w'\theta'}$ may also have negative values. The ratio of buoyancy-over shear production (the first two terms in (29)) is known as the Richardson number,

$$Ri = -\frac{g}{T}\frac{\overline{w'\theta'}}{\overline{u'w'}\partial U/\partial z}. \tag{30}$$

When $Ri > 0$, the negative heat flux suppresses the atmospheric turbulence, a situation which is typical for a stable (night time) atmospheric boundary layer (cf. the qualitative discussion in the section "Thermodynamics"). When $Ri < 0$ turbulence is generally enhanced, a situation typical for the daytime, unstable surface layer.

Another stability parameter, the Obukhov length, L, may be obtained from (29) as the height at which the buoyancy production equals the shear production. Approximating $-\overline{u'w'}\partial U/\partial z$ by u_*^3/kz, we have

$$\left(\frac{u_*^3}{kz}\right)_{z=L} = -\frac{g}{T}\overline{w'\theta'}$$

whence,

$$L = -\frac{u_*^3}{k(g/T)\overline{w'\theta'}}. \tag{31}$$

Finally we introduce another measure of stability, which is related to the oscillatory behaviour of air parcels in a stable atmosphere (cf. section "Thermodynamics"). It can easily be derived that the harmonic force in a stable atmosphere $(d\theta/dz > 0)$ is proportional to the lapse rate and the acceleration of gravity. This defines the Brunt Väisälä frequency, N by

$$N = \left[\frac{g}{T}\frac{d\Theta}{dz}\right]^{1/2}. \tag{32}$$

So far we have shown that in the neutral surface layer $(z \gg z_0$, but $z \ll h$, the boundary layer height) the wind profile is logarithmic, or

$$\left(\frac{z}{u_*}\right)\frac{\partial U}{\partial z} = \text{constant} \ (= 1/k)$$

(cf. (27, 28)). In a stratified surface layer $(L^{-1} \neq 0)$ the wind profile is different from (27, 28). Dimensional analysis suggests that the only available parameters in the stratified surface layer are z and L. Hence

$$(kz/u_*)\frac{\partial U}{\partial z} = \phi_m(z/L), \tag{33}$$

where ϕ_m is a function of the dimensionless parameter z/L. Similarly for the temperature profile

$$(kz/\theta_*)\frac{\partial\Theta}{\partial z} = \phi_h(z/L). \tag{34}$$

where θ_* is a temperature scale in the surface layer, defined by

$$\theta_* u_* = -(\overline{w'\theta'})_0. \tag{35}$$

The functions ϕ_m and ϕ_h were experimentally established [5]. Upon integration (33) and (34) yield

$$U = (u_*/k)\{\ln(z/z_0) - \psi_1(z/L)\} \quad \text{and} \tag{36}$$

$$\Theta = \Theta_0 + 0.74(\theta_*/k)\{\ln(z/z_0) - \psi_2(z/L)\}, \tag{37}$$

where for $L < 0$

$$\psi_1 = 2\ln\left(\frac{1+x}{2}\right) + \ln\left(\frac{1+x^2}{2}\right) - \arctan(x) + \frac{1}{2}\pi; \quad x = \left(1 - 15\frac{z}{L}\right)^{1/4}$$

$$\tag{39}$$

$$\psi_2 = 2\ln\left(\frac{1+y}{2}\right); \quad y = \left(1 - 9\frac{z}{L}\right)^{1/2},$$

and for $L > 0$

$$\psi_1 = -4.7z/L \quad \text{and} \quad \psi_2 = -6.4z/L.$$

Because there are certain similarities between the turbulent transport of heat and a passive contaminant, (37) is often adopted to describe the contaminant concentration in the surface layer.

When we rewrite the eddy coefficient K defined in the beginning of section 3.2 as $K_m = -\overline{u'w'} \left(\dfrac{\partial U}{\partial z}\right)^{-1}$, we may derive from the expression for $\overline{u'w'}$ and $(\partial U/\partial z)^{-1}$ in the neutral surface layer (26) and (27) the formula

$$K_m = kzu_* . \tag{40}$$

This can be generalised to the stratified case by putting

$$K_m = kzu_*/\phi_m(z/L)$$

and similarly

$$\tag{41}$$

$$K_h = kzu_*/\phi_h(z/L).$$

where the indices refer respectively to the exchange coefficients of momentum and heat.

The Convective Boundary Layer

So far we have only discussed the mean flow and turbulence characteristics in the surface layer, a thin layer, which extends to an altitude of only a few tens of metres generally. During daytime when the insolation of the surface is strong enough to induce convective thermal motions, a mixed layer develops with an approximately constant potential temperature. On sunny days the mixed-layer height gradually increases to 1000–2000 m in the late afternoon. In Fig. 5 a schematic picture is

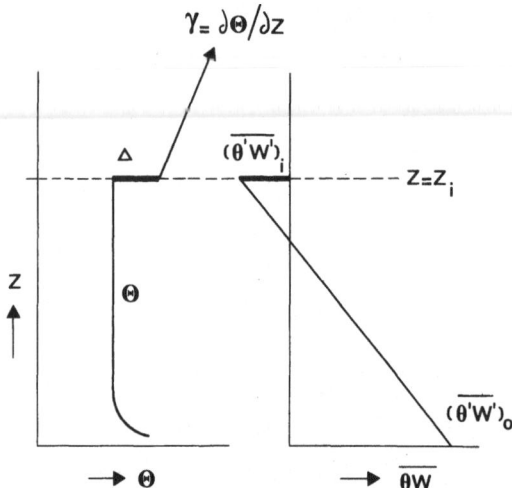

Fig. 5. Schematic representation of the potential temperature profile and heat flux in a convective boundary layer (CBL). In the major part of the boundary layer the temperature profile is neutral ($d\theta/dz = 0$). The CBL is capped by a shallow inversion at $z = z_i$. Turbulent motions at $z = z_i$ entrain continuously stable air ($d\theta/dz > 0$), resulting in a gradual increase of the boundary layer height during morning and afternoon [6].

given of the vertical temperature profile and vertical heat flux. For this "model" of the mixed layer a dynamical model was developed [6] which could describe the evolution in time of the mixed-layer height, z_i. For simplicity we give an approximate relationship only [7], which reads in differential form:

$$\frac{dz_i}{dt} = \frac{(\overline{\theta' w'})_0}{\gamma z_i (1 - c_3 L/z_i)} [1 + c_1 + c_2 (L/z_i)^2], \tag{42}$$

where γ is the lapse rate $(d\theta/dz)$ of the layer above the mixed layer. The numerical constants $c_1 - c_3$ are 0.4, 31, and 8.7 respectively. Given an initial value for z_i, the mixed layer height can be obtained from (42). The height of the mixed layer is one of the most important parameters for the dispersion of air pollutants. Note that (42) was derived for a boundary layer which is horizontally homogeneous.

The Stable Boundary Layer

The radiative cooling, generally during night time, results in a gradual fall of surface temperature, which is dispersed into the adjacent atmospheric layer. This results in the build-up of a stable nocturnal boundary layer $(d\theta/dz > 0)$. The growth of this layer is less rapid than that of the convective boundary layer, because (vertical) turbulent motions, the main transport mechanism of heat, are much less vigorous during the night. Initially there is a rapid growth, which seems to level off towards an equilibrium value. A convenient expression for this value is implicitly given by the equation [8]

$$z_i = c \frac{u_*/f}{1 + 1.9 z_i/L}, \tag{43}$$

where the constant of proportionality is 0.2 (the more common value 0.3 leads to unrealistically high values of z_i). In near neutral conditions (43) reduces to

$$z_i = c u_*/f, \tag{44}$$

a relation which is in qualitative agreement with the neutral boundary layer height, h, when for $K = k h u_*$ (40), is substituted in (23). The wind velocity in the stable boundary layer changes normally strongly with height. Sometimes local windmaxima occur near or just below the inversion- or boundary layer height (nocturnal or low level jet). Above the stable boundary layer the wind velocity is well represented by the geostrophic relationship (21).

In the convective boundary layer the wind varies with height in the thin surface layer only. The strong and continuous vertical motions in the mixed layer inhibit the existence of strong gradients in the (mean) wind velocity. Near the inversion height there may be a jump in wind velocity, when the wind adopts the geostrophic value (21).

Transport and Dispersion

The mathematical treatment of the dispersion of air pollution in a turbulent flow starts from the equation of conservation of mass. For example, consider a parcel of

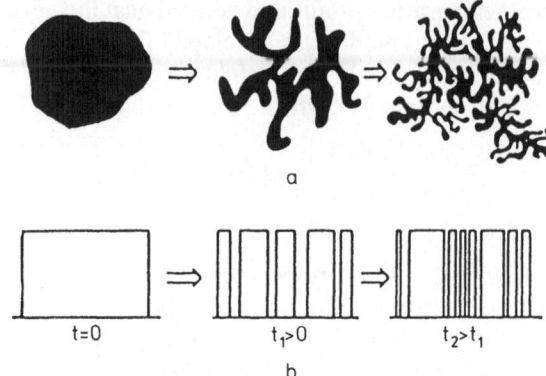

Fig. 6. (a) An illustration of the dispersion of a "blob" of marked turbulent fluid (Molecular diffusion is neglected). Turbulent motions on various scales distort and strain the original parcel. (b) Below the concentration c is schematically indicated, which shows that the measuring probe is with increasing intermittency within or without the marked fluid. (Source: Monin and Yaglom I, MIT 1971)

air containing "marked particles" with concentration c. (It is assumed that the marking of the particles has no influence on the dynamics of the flow). This parcel is, neglecting molecular diffusion (or Brownian motion when the marked particles are real particles (smoke)), moved around in the surrounding air, becoming more and more distorted by the turbulent flow (Fig. 6), however, without a change in its concentration, dc/dt = 0. Noting that in a fixed frame or reference (Eulerian description) c depends on the coordinates (x, y, z) and time (t), we have

$$\frac{\partial c}{\partial t} + u\,\frac{\partial c}{\partial x} + v\,\frac{\partial c}{\partial y} + w\,\frac{\partial c}{\partial z} = 0, \tag{45}$$

or, using the incompressibility condition (18)

$$\frac{\partial c}{\partial t} + \frac{\partial uc}{\partial x} + \frac{\partial vc}{\partial y} + \frac{\partial wc}{\partial z} = 0. \tag{46}$$

Here also we decompose the velocity and concentration into a mean and a fluctuating component. After some algebra and averaging we obtain

$$\frac{\partial C}{\partial t} + U\,\frac{\partial C}{\partial x} + V\,\frac{\partial C}{\partial y} + W\,\frac{\partial C}{\partial z} = -\left(\frac{\partial \overline{u'c'}}{\partial x} + \frac{\partial \overline{v'c'}}{\partial y} + \frac{\partial \overline{w'c'}}{\partial z}\right), \tag{47}$$

the basic transport equation for the dispersion in a turbulent flow. Mean quantities are expressed by capital symbols. The fluctuating quantities are indicated by primes.

The "gradient transfer" method suggests to put

$$\overline{u'c'} = -K_x \partial C/\partial x \quad \text{etc.} \tag{48}$$

(More generally $\overline{u'c'} = -K_{xj}\partial C/\partial x_j$ etc., where the repeated index j implies summation over the three coordinates and K_{ij} is a diffusivity tensor. Here we assume that only the diagonal elements $K_{xx} = K_x$ etc. are non-zero). Hence

$$\frac{\partial C}{\partial t} + U\frac{\partial C}{\partial x} + V\frac{\partial C}{\partial y} + W\frac{\partial C}{\partial z}$$

$$= \frac{\partial}{\partial x}\left(K_x\frac{\partial C}{\partial x}\right) + \frac{\partial}{\partial y}\left(K_y\frac{\partial C}{\partial y}\right) + \frac{\partial}{\partial z}\left(K_z\frac{\partial C}{\partial z}\right). \tag{49}$$

Equation (49) is the basic transport equation in turbulent flow. Because it is presented in the Eulerian framework it is referred to as the Eulerian transport equation.

Analytical Solutions

In the kinematic approach it is assumed that the (mean) wind field and eddy diffusivities are given as function of space and time. Then (49) can be solved with adequate initial- and boundary conditions. The complexity of the differential equation, however, inhibits analytical solutions in most cases of practical interest.

In a few simple cases an analytical solution can be obtained, which requires, however, some further simplifications of (49). First we assume that the wind field is steady and uniform and that average vertical motions may be neglected ($W=0$). Further along wind diffusion is neglected ($K_x\partial C/\partial x = 0$) and it is assumed that the lateral diffusion coefficients are constants. The stationary form of Eq. (49) then reduces to

$$U\frac{\partial C}{\partial x} = K_y\frac{\partial^2 C}{\partial y^2} + K_z\frac{\partial^2 C}{\partial z^2}. \tag{50}$$

When we introduce the boundary conditions

$$C\to 0 \quad \text{for} \quad y\to \pm\infty \quad \text{and} \quad z\to\infty \tag{51}$$

and

$$K_z\frac{\partial C}{\partial z} \to 0 \quad \text{for} \quad z\to 0,$$

we are able to obtain an analytical solution, which describes the dispersion from a point source at the position $(0, 0, H)$. Thus when initially

$$C(x=0, y=0, z=H) = \frac{Q}{U}\,\delta(z-H)\,\delta(y) \tag{52}$$

the solution of (50) is

$$C(x, y, z) = \frac{Q}{2\pi\sigma_y\sigma_z U}\, e^{-\frac{1}{2}(y/\sigma_y)^2}$$

$$\cdot \{e^{-\frac{1}{2}[(z-H)/\sigma_z]^2} + e^{-\frac{1}{2}[(z+H)/\sigma_z]^2}\}, \tag{53}$$

the well-known Gaussian plume solution. Here, $\sigma_{y,z}$ are the horizontal and vertical dispersion coefficients defined by

$$\sigma_{y,z}^2 = 2K_{y,z}x/U. \tag{54}$$

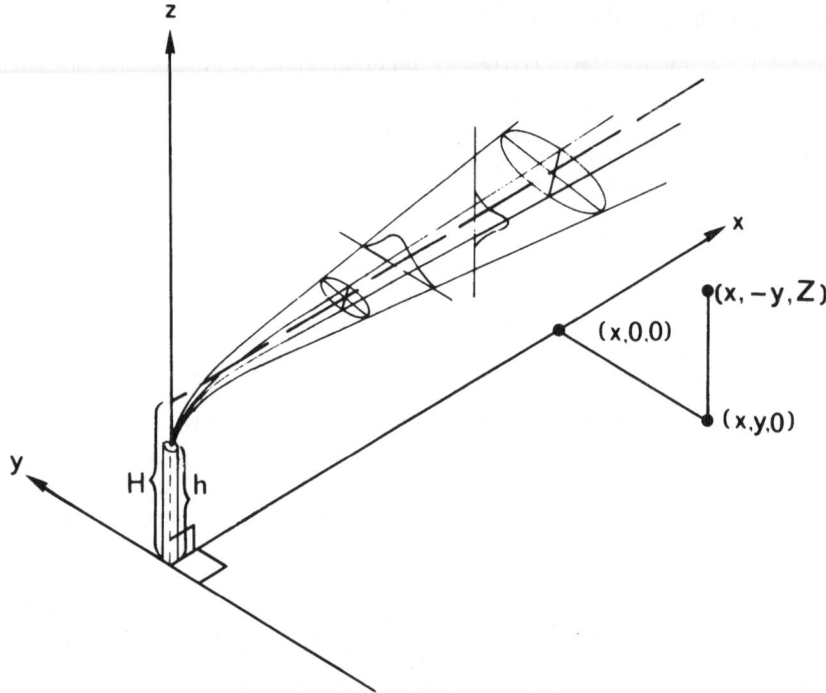

Fig. 7. Sketch of Gaussian dispersion from a point source. The difference between H and h (stack height) is the plume rise Δh. (Source: luchtverontreiniging en weer, Staatsuitgeverij, Den Haag 1979)

Equation (53) shows that in ideal conditions the pollution, which is released for example from a chimney, spreads into Gaussian shaped concentration distributions, centred around the release point (Fig. 7). The use of (53) is widespread notwithstanding the severe limitations in its derivation. It is usually applied with dispersion coefficients, based on empirical data. Their x-dependance is usually different from (54), as we shall see in the section "Dispersion Coefficients."

Another interesting case is obtained by considering the cross-wind integrated concentration,

$$C_y \equiv \int\limits_{-\infty}^{+\infty} C \, dy.$$

Integrating (50) over y yields the equation

$$U \frac{\partial C_y}{\partial x} = \frac{\partial}{\partial z}\left(K_z \frac{\partial C_y}{\partial z} \right), \tag{55}$$

for which analytical solutions can be obtained when U and K_z depend on z according to power laws,

$$U(z) = U_1 \left(\frac{z}{z_1} \right)^m \quad \text{and} \quad K_z(z) = K_1 \left(\frac{z}{z_1} \right)^n. \tag{56}$$

Using similar boundary conditions for z as in the previous case (51) we obtain

$$C_y = \frac{Qr}{2z_1 U_1 \Gamma(s)} \left[\frac{z_1^2 U_1}{r^2 K_1 x} \right]^s \cdot \exp \left[-\left(\frac{z}{z_1} \right)^r \frac{U_1 z_1^2}{r^2 K_1 x} \right], \tag{57}$$

with $r \, (= m - n + 2) > 0$, and $s = (m + 1)/r$.

Equation (57) reduces to the Gaussian form for $m, n = 0$. There is a broader range of wind- and diffusivity profiles for which analytical solutions may be obtained. They can be found in, e.g., ref. [1, 4, 9] and [10].

Statistical Dispersion Theory

In the previous section dispersion of material is described by a conservation equation in the Eulerian framework. In order to solve the equation a closure assumption (48) is required, which puts some limitations on the validity of the solutions. Generally, one might say that (48) is only valid when the spatial scale of the diffusing motions, the turbulent eddies, is much smaller than the characteristic scale of the dispersing plume (as is, e.g., the case in molecular diffusion). Also the parabolic character of the resulting diffusion equations implies that the spreading of the plume is immediate, which suggests that diffusion takes place with infinite velocity [11]. These objections make the application of (49) etc. often doubtful, certainly when one considers diffusion close to a continuous source, (or, equivalently, the diffusion of a instantaneously released puff shortly after its time of release).

An approach which avoids these objections is the statistical theory of dispersion, originally introduced by G.I. Taylor [12]. It is formulated in a Lagrangian framework, i.e., the displacement of individual, marked (polluted) parcels is studied, by determining their trajectory. Then, for example, by considering many trajectories with a common starting point (the source) one may obtain statistical properties such as mean particle position, variance etc., by taking an adequate average over a sufficient amount of particles. When we denote the marked particle position by the capital symbols $X(t)$, $Y(t)$, $Z(t)$ and the corresponding velocity components by $U(t)$, $V(t)$, and $W(t)$, noting that $U(t) \equiv dX(t)/dt$ etc., we may write for the vertical particle position, $Z(t)$,

$$Z(t) = \int_0^t W(t') dt', \tag{58}$$

with the implicit assumption that the particle left the origin at $t = 0$. Additionally it is assumed that at $t = 0$ the particles have zero mean velocity and that their velocity variance equals the fluid velocity variance σ_w^2. From (58) we have also

$$\frac{dZ^2}{dt} = 2Z \frac{dZ}{dt} = 2 \int_0^t W(t') W(t) dt'. \tag{59}$$

Now we take the average of the ensemble of particles which at $t = 0$ all passed through the origin (Fig. 8)

$$\frac{d\overline{Z^2}}{dt} = 2 \int_0^t \overline{W(t') W(t)} dt', \tag{60}$$

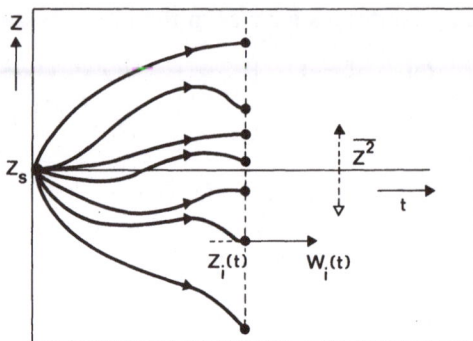

Fig. 8. Dispersion of a few marked particles in one dimension. $\left(Z_i(t) \text{ and } W_i(t) \equiv \dfrac{dZ_i}{dt} \text{ denote the} \right.$ position and velocity of the i-th particle respectively $\left. \vphantom{\dfrac{dZ_i}{dt}} \right)$. The particle spread $\overline{z^2} \equiv \dfrac{1}{n} \sum\limits_{i=1}^{n} (z_i - \bar{z})^2$ becomes identical to (61) when n is large enough. (The mean particle position $\bar{Z}(t)$ equals zero (take $z_s = 0$) in homogeneous turbulence).

hence,

$$\overline{Z^2} = 2 \int\limits_0^t dt'' \int\limits_0^{t''} \overline{W(t') W(t'')} \, dt', \tag{61}$$

which relates the particle position variance to the Lagrangian velocity. Introducing the correlation function $R_L(t', t'')$ by

$$R_L(t', t'') = \overline{W(t') W(t'')} / [\overline{W(t')^2} \ \overline{W(t'')^2}]^{1/2} \tag{62}$$

we may rewrite (61) in a homogeneous and stationary turbulence, where $\overline{W(t')^2} = \overline{W(t'')^2} \equiv \sigma_w^2$ and $R_L(t', t'') = R_L(\|t' - t''\|)$ as

$$\overline{Z^2} = 2\sigma_w^2 \int\limits_0^t dt' \int\limits_0^{t'} R_L(\tau) d\tau. \tag{63}$$

Though (63) gives a more physical description of the diffusion problem than the gradient-transfer diffusion equation (55), it leaves us with the problem to determine $R_L(\tau)$. This requires the keeping track of individual fluid particles in a turbulent flow, in general a difficult experimental task. Some general considerations, however, may already give some insight in the diffusion behaviour: close to the source, near $t = 0$, the inter-particle distance is small and $R_L(\tau)$ is close to unity. Then (63) reduces to

$$\overline{Z^2} = \sigma_w^2 t^2. \tag{64}$$

Generally $R_L(\tau)$ drops to zero when τ becomes larger than the timescale, T_L, of the turbulent flow and fluid particles tend to move independently from each other. Changing the order of integration in (63) yields

$$\overline{Z^2} = 2\sigma_w^2 \int\limits_0^t (t - \tau) R_L(\tau) d\tau, \tag{65}$$

which for $t/T_L \gg 1$ is approximately equal to

$$\overline{Z^2} = 2\sigma_w^2 \left\{ \int_0^t R_L(\tau)d\tau \right\} t. \tag{66}$$

Noting that then $\int_0^t R_L(\tau)d\tau$ is virtually equal to $\int_0^\infty R_L(\tau)d\tau \equiv T_L$, (66) reduces to

$$\overline{Z^2} = 2\sigma_w^2 T_L t, \tag{67}$$

when $t/T_L \gg 1$.

Thus the assumption that $R_L(\tau)$ is only significantly different from zero for a range of time which is limited by the (Lagrangian) time scale T_L, yields the results (64, 67). They indicate that initially the spreading of an ensemble of particles is linear with t and at large time is proportional to $t^{1/2}$.

When the one dimensional equivalent of the eddy diffusion equation for a flow with zero mean velocity (49),

$$\frac{\partial C}{\partial t} = \frac{\partial}{\partial z} \left(K_z \frac{\partial C}{\partial z} \right) \tag{68}$$

is multiplied with z^2 and integrated over z we obtain, noting that $\overline{Z^2} \equiv \int z^2 C(z)dz / \int C(z)dz$,

$$\frac{d\overline{Z^2}}{dt} = 2K_z. \tag{69}$$

Comparing (69) and (67) we may put

$$K_z = \sigma_w^2 T_L. \tag{70}$$

Assuming that in homogeneous turbulence the probability distribution of the particle displacement is approximately normal we observe that the statistical description of turbulent dispersion and the gradient-transfer approach (68) are equivalent when $t \gg T_L$. Conversely, (68) is not valid for small times.

Finally, a convenient expression for $R_L(\tau)$ with the above properties is

$$R(\tau) = e^{-t/T_L} \tag{71}$$

which, introduced in (65) yields

$$\overline{Z^2} = 2\sigma_w^2 T_L^2 \left(\frac{t}{T_L} - 1 + e^{-t/T_L} \right), \tag{72}$$

already obtained by G. I. Taylor [12] in 1921.

Random Walk Dispersion Theory

There is a renewed interest in the study of dispersion by means of stochastic, or random walk methods. It is assumed that the movements of parcels of fluid is governed by subsequent, independent random accelerations, much like Brownian motion [13]. The acceleration of a fluid particle in homogeneous and steady turbulence can be written as

$$\frac{dW}{dt} = -\alpha W + \beta \omega(t), \tag{73}$$

where α and β are coefficients to be determined by the physical properties of the fluid flow. The function $\omega(t)$ is a stochastic function, which means that only its statistical properties are known. Usually it is assumed that $\omega(t)$ is Gaussian and $\overline{\omega(t)}=0$, $\overline{\omega(t)\omega(t')}\simeq 0$ for $\|t-t'\|$ larger than a very small time, t_k, $(t_k \ll T_L)$ in accordance with the assumption that subsequent accelerations are independent. The coefficient α in (73) indicates the time range over which $W(t)$ is correlated and can – in view of the previous section be identified with $1/T_L$. The requirement that the velocity of the marked fluid particles become asymptotically equal to the (constant) fluid velocity variance, σ_w^2, determines the coefficient β in (73):

$$\beta = \sigma_w (2/T_L)^{1/2} \tag{74}$$

Using $dZ = W dt$, we may integrate (73) twice, average and obtain an expression for $\overline{Z^2}$ which is identical to that of Taylor [12] (72).

For the application in random walk or Monte Carlo models (73) is often written in a forward explicit finite difference form

$$W(t+\Delta t) = W(t)\left(1 - \frac{\Delta t}{T_L}\right) + \sigma_w \left(\frac{2\Delta t}{T_L}\right)^{1/2} n$$

and $\hspace{11cm} (75)$

$$Z(t+\Delta t) = Z(t) + W(t)\Delta t,$$

where n is a random number, which is drawn from a Gaussian distribution with mean zero and unit variance ($\bar{n}=0$ and $\overline{n^2}=1$). With (75) we are able to construct random walks or trajectories of fluid particles. It can be shown [14] that (75) also works in a turbulence where T_L is a function of the particle position (Z), (for example in a neutral turbulent boundary layer). In non-homogeneous or non-stationary turbulence (75) is no longer valid. In that case more complicated random walks models should be used [15–17].

Numerical Methods

It has already been stated that the solution of the general transport-diffusion equation (49) requires numerical methods. A much used procedure is the method of fractional steps, which consists here of splitting the equation in an advection part and a diffusion part. Both parts are subsequently and alternatively evaluated in time. Thus one first advects the pollutant field according to

$$\frac{\partial C}{\partial t} + U\frac{\partial C}{\partial x} + V\frac{\partial C}{\partial y} + W\frac{\partial C}{\partial z} = 0 \tag{76}$$

and then diffuses the field according to

$$\frac{\partial C}{\partial t} = \frac{\partial}{\partial z}\left(K_z \frac{\partial C}{\partial z}\right) + S \tag{77}$$

where for simplicity the diffusion is considered in only one (vertical) direction. All sources and sinks are contained in S. Equations (76, 77) can now be solved with standard numerical techniques which are appropriate for the respective form of the differential equation. The first step in the solution is writing the equations in finite difference form, e.g.,

$$\frac{\partial C}{\partial t} = \frac{C^{n+1} - C^n}{\varDelta t} + O(\varDelta t)$$

$$\frac{\partial C}{\partial x} = \frac{C_{i+1} - C_{i-1}}{2\varDelta x} + O(\varDelta x^2),$$

$$(78)$$

where the upper indices denote the (discrete) time and the lower indices the (discrete) space variables. Note that (78) is to first order in $\varDelta t$ and to second order in $\varDelta x$. Using (78), we may write the one dimensional advection equation, $\partial C/\partial t + U \partial C/\partial x = 0$ as

$$C_i^{n+1} = C_i^n - \tfrac{1}{2}\lambda(C_{i+1}^n - C_{i-1}^n) + O(\varDelta t \varDelta x^2), \qquad (79)$$

where λ, the Courant number, is $U\varDelta t/\varDelta x$, and should be chosen small enough to obtain the desired accuracy and stability of the solution. Generally the problem of solving the differential equation is reduced to solving a matrix equation, for which standard numerical techniques exist [18].

The above finite difference scheme is an explicit Euler scheme and is always unstable. The accuracy and stability of finite difference schemes can be improved by using higher order expansions in time and space variables. In Table 3 a listing is given of some current finite different schemes for the advection equation. The pseudo spectral method has gained some popularity with the advent of fast numerical Fourier transform techniques. In this case the advection equation is

Table 3. Some current numerical advection schemes

Method	Description	Reference
Gadd	4-th order extension of Lax-Wendroff scheme	A.J. Gadd [43]
Praagman	4-th order centered difference in space plus Runge Kutta 4-th order in time (RK4)	N. Praagman [44]
Fox-Orszag	Pseudo-spectral method in combination with fast Fourier transform (RK4 in time)	Fox and Orszag [45]
PS/L	Pseudo spectral method; leap-frog scheme in time	Chock and Dunker [46]
CF/I	Chapean-function method; implicit Crank Nicolson in time	Pepper et al. [47, 48]
CF/FED	Linear finite element method; forward Euler in time with balanced diffusion	Chock [48]

solved in Fourier space by the introduction of

$$\tilde{C}(k, t) = \int e^{-ikx} C(x, t) dx,$$

so that the one-dimensional advection equation can be written as

$$\frac{\partial \tilde{C}}{\partial t} - iUk\tilde{C} = 0, \tag{80}$$

which is an ordinary first order differential equation in time and easily can be solved. Though the method is attractive since a coarser grid (less grid points) can be used to obtain the same accuracy as in conventional finite differential schemes, it requires careful treatment of advection at the boundaries [19, 20].

The solution of parabolic type differential equations (77) is usually obtained by the finite difference form

$$\frac{C_j^{n+1} - C_j}{\Delta t} = \frac{K_z}{\Delta z^2} \{\theta(\delta^2 C)_j^{n+1} + (1-\theta)(\delta^2 C)_j^n\}, \tag{81}$$

where $\delta C = C_{j+\frac{1}{2}} - C_{j-\frac{1}{2}}$ and θ may be chosen for example 0 (Euler explicit), $\frac{1}{2}$ (Crank-Nicolson), 1 (fully implicit), or $\frac{1}{2} - \Delta z^2/(12K_z\Delta t)$.

If, apart from the pollutant transport, also chemical reactions play a role additional non-linear equations have to be solved of the type

$$\frac{\partial C_i}{\partial t} = \sum_{j,l} k_{ijl} C_j C_l, \tag{82}$$

where the concentration of species i is indicated by C_i. For the solution of (82) special routines exist which can cope with the problem that the reaction coefficients k_{ijl} may vary over a few orders of magnitude, since both slow and fast reaction are involved (the "stiffness problem") [21].

Present computer size is such that usually a numerical method can be selected with the desired accuracy and efficiency. Which particular scheme should be used cannot be prescribed in general. It depends on the particular application and the available computer facilities.

Air Pollution Models

It is useful to distinguish between various spatial- and corresponding time scales in air pollution modelling. Often processes, which may be neglected when dispersion is considered over some tens of km only, may be important when larger distances (and times) are involved. On the other hand processes, which govern the dispersion on a small scale, may be ignored when studying dispersion on, e.g., the global scale. An example of the former is dry deposition, while in the latter case one may think of plume rise or the daily variation in atmospheric stability.

Dispersion on the following scales is therefore separately treated:
1) Urban dispersion models, which operate on a spatial scale of approximately 0–50 km, corresponding with a dispersion time of $\sim$ one hour.

2) Medium range or meso scale models, operating on a spatial scale up to ca. 500 km, describing transport over 1–2 days.
3) Long range transport models, which operate on the continental or global scale.

Urban Dispersion Models

The scale of 0–50 km, corresponding with a dispersion time of ~ 1 hour is more or less representative for an urban or industrial area. Meteorological conditions are assumed quasi steady during one hour: mean wind velocity and turbulent is assumed constant. Also spatial uniformity is assumed. Then for a point source the concentration distribution is given by (53). This distribution is of course a strong idealisation due to the large amount of assumptions made. However, using (53) one may obtain more or less reliable statistical concentration distributions, when an average is taken over a longer period of time, say from one month or longer.

The variables in (53) are the dispersion coefficients (σ_y, σ_z) and the mean wind velocity (U). So when for a particular site a frequency distribution (or climatology) of these parameters can be determined, we may determine the average concentration over a fixed period of time. Dispersion coefficients are related to the eddy diffusivity (cf. (54)) and therefore to the turbulent state of the atmosphere, its stability. (See Table 4). To each stability class a particular dispersion coefficient is assigned.

Table 4. Horizontal and vertical dispersion coefficient estimates. Their dependence on the distance is given according to $\sigma_{y,z} = ax^s$. Pasquill's values pertain to surface releases (a) and Smith's to elevated releases (b) (σ and x are in km). Table 4b is based on measurements over terrain with $z_0 \simeq 1$ m. (Source: Pasquill and Smith [1])

(a)

Stability	Coefficient a			Index s		
z_0	1 cm	10 cm	1 m	1 cm	10 cm	1 m
A	0.102	0.140	0.190	0.94	0.90	0.83
B	0.062	0.080	0.110	0.89	0.85	0.77
C	0.043	0.056	0.077	0.85	0.80	0.72
D	0.029	0.038	0.050	0.81	0.76	0.68
E	0.017	0.023	0.031	0.78	0.73	0.65
F	0.009	0.012	0.017	0.72	0.67	0.58

(b)

Stability	σ_y		σ_z	
	Coefficient a	Index s	Coefficient a	Index s
Very unstable	0.215	0.91	0.215	0.91
Unstable	0.137	0.86	0.125	0.86
Neutral	0.070	0.76	0.048	0.76
Stable	0.042	0.71	0.008	0.71

The distribution of wind velocity is made by dividing wind speed and direction in a number of classes, N, e.g., three wind speed classes (0.5–2.5, 3–5.5 and $\geq 6\,\mathrm{m\,s^{-1}}$) and 12 wind direction sectors of 30° width. It is assumed that the hourly plume of a single point source is always within the breadth of one wind direction sector ("slender plume approximation") and for simplicity the concentration distribution is taken uniform in the y-direction within each 30° wind direction sector. In other words, in (53) $1/((2\pi)^{1/2}\sigma_y)\exp[-\tfrac{1}{2}(y/\sigma_y)^2]$ is replaced by $1/((2\pi/n)x)$ where n is the number of wind direction sectors, so that (53) becomes

$$C_{\theta,N,S}(x,z)=\frac{Q}{(2\pi)^{1/2}U_n\sigma_{z_s}(2\pi x/n)}\left\{e^{-\frac{1}{2}\left(\frac{z+H}{\sigma_{z_s}}\right)^2}+e^{-\frac{1}{2}\left(\frac{z+H}{\sigma_{z_s}}\right)^2}\right\}\qquad(83)$$

where θ refers to the particular wind direction considered ($\theta=1,...,12$), and the dispersion coefficients are labelled according to their stability class ($S=1,...,6$). The long term average concentration is obtained by summing (83) over all wind directions, wind speeds and stability classes with the appropriate weighting function according to their frequency of occurrence. Hence from (83)

$$C_{av}=\sum_{\theta,N,S}f(\theta,N,S)C_{\theta,N,S}(x,z),\qquad(84)$$

where $f(\theta,N,S)$ denotes the weighting function.

An example of $f(\theta,N,S)$ is given in Table 5, where statistical data are compiled for Schiphol [22] (Amsterdam airport), over the years 1949–1970.

Equation (84) determines the average concentration distribution for a single source, with source strength Q. By repeating the calculation for each source which is considered to give a significant contribution and adding the concentration distributions we obtain the final average concentration distribution in the region.

Table 5. Climatology of Schiphol $f(\theta, N,S)$ based on hourly observations during 1948–1970. (Total amount of observations 179112). The wind direction classes, θ, involve sectors of 30° such that $\theta=0$ indicates wind direction between $345°-15°$; $\theta=1$, $15°-45°$ etc. The stability class (S) and wind speed class (N) are defined in the text

S	A	A	B	B	C	C	C	D	D	D	E	E	F	F
N	1	2	1	2	1	2	3	1	2	3	1	2	1	2
θ														
0	170	30	234	533	40	470	352	1075	2761	2984	175	601	1260	271
1	153	26	161	624	25	483	630	744	2082	2872	99	536	689	181
2	200	44	206	1209	45	688	1214	986	3514	5616	93	909	866	345
3	279	42	239	990	40	461	441	1168	3641	3497	133	1310	1022	372
4	202	19	211	553	34	300	147	1028	2490	1372	145	620	879	221
5	195	27	217	521	48	367	101	1350	4070	3123	212	984	984	318
6	223	28	302	514	55	491	117	1766	5586	5617	319	973	1284	338
7	166	23	218	465	42	593	164	1476	6356	11480	240	1112	822	304
8	180	29	226	487	33	584	555	1507	4456	13153	259	780	1093	266
9	186	30	258	757	32	780	536	1286	3661	9744	169	520	809	174
10	146	24	196	575	36	572	285	989	2584	6063	148	397	664	125
11	165	36	208	535	46	571	309	1082	2817	4539	160	495	983	169

Dispersion Coefficients

It is common practice to use in (83) empirical dispersion coefficients. They are based on some classical experiments where plume widths were measured in well-defined meteorological conditions. These conditions were subdivided into the same classes of wind speed and stability as above. It was assumed that the dispersion had universal validity, provided that the terrain conditions were comparable. However, a simple procedure was developed to convert the dispersion coefficients to terrain types of different roughness.

Because turbulence and consequently dispersion varies with altitude, a rough distinction is made for the release from "low" and "high" sources. In Fig. 9 and 10 dispersion coefficients are presented according to Pasquill. For high sources the dispersion coefficients based on the field experiments by Singer and Smith (Table 4) are usually adopted. There are, however, many other data on dispersion available [10]. Recently more schemes became available, which could incorporate the scaling parameters of the section "turbulence" (cf. Eqs. (26, 30, 31, 32)) [32].

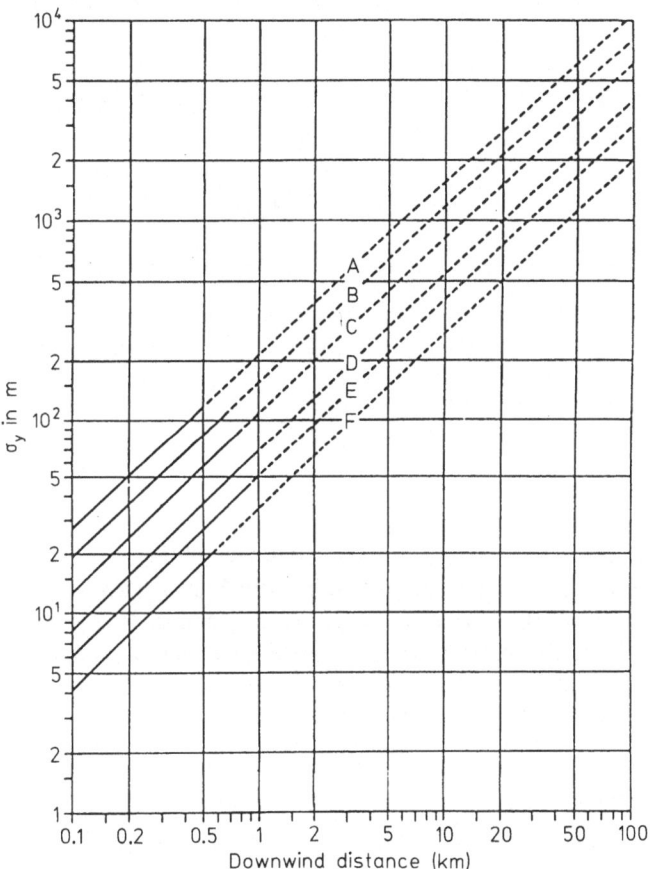

Fig. 9. Horizontal dispersion coefficient, σ_y, according to Pasquill as a function of the distance from the source for 6 stability categories, A–F. (Source: luchtverontreiniging en weer, Staatsuitgeverij, Den Haag 1979)

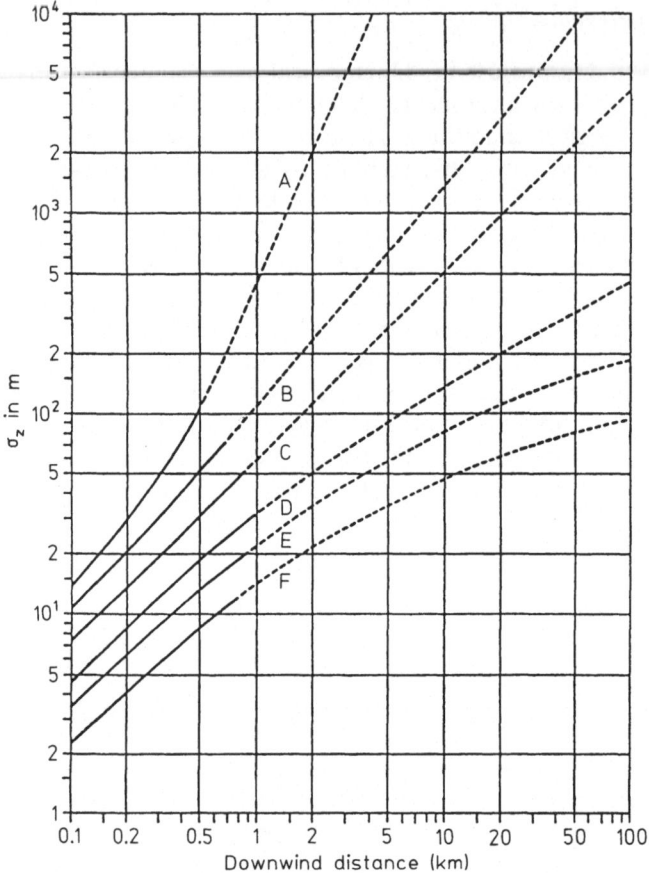

Fig. 10. As Fig. 9 but now for the vertical dispersion coefficient, σ_z. (Source: luchtverontreiniging en weer, Staatsuitgeverij, Den Haag 1979)

Plume Rise

Gaseous material released from an industrial source is often much hotter than the ambient air and buoyancy forces will cause an ascent of the plume to an altitude which often exceeds the chimney height significantly. During its ascent the plume is gradually bent towards the horizontal mean wind direction. It is clear that in (83) this final height should be used for H and not the stack height h (cf. Fig. 7). Briggs [23] has formulated some expression for the plume rise, Δh. Introducing the stack parameter $F = gQ_H/(\pi \varrho c_p T_0)$, where Q_H is the heat output of the source (Watt) and T_0 the ambient temperature, we have

$$\Delta h = F^{1/3} U^{-1} x^{2/3} \quad \text{in neutral conditions} \tag{85}$$

$$= F^{1/3} U^{-1/3} N^{-2/3} \quad \text{in stable conditions (final rise)} \tag{86}$$

$$= F^{1/4} N^{-3/4} \quad \text{in stable, calm conditions (final rise),} \tag{87}$$

where N, the Brunt-Vaisala frequency, is the parameter which determines the stability of the ambient air (cf. Eq. (32)).

Air Pollution Statistics

With (84) an average concentration field can be evaluated. However, for health and safety reasons one is often interested in the occurrence of peak concentrations and therefore (83) can be used to assemble a frequency distribution of concentrations at a given site. This should be done with considerable care because the abundant amount of model assumptions induces errors which may lead to large inaccuracies in the prediction of high percentile values, because these values correspond with meteorological circumstances which seldomly occur. (Note that an n-percentile value of a concentration distribution is by definition that concentration value above which n% of all the observations are. Applied to a concentration distribution over a year, which is built up from daily mean values for example, the 2-percentile is the seventh highest concentration value).

Larsen [24] developed an empirical method to determine high percentile values. Based on observations in large urbanised areas he gives a relation between the average concentration and percentile values. An example of his results is summarised in Fig. 11.

The Validation of Urban Air Pollution Models

A variety of models based on (53) or (83) have been developed for many kinds of applications [1, 26]. Numerous modifications have been proposed, many of which

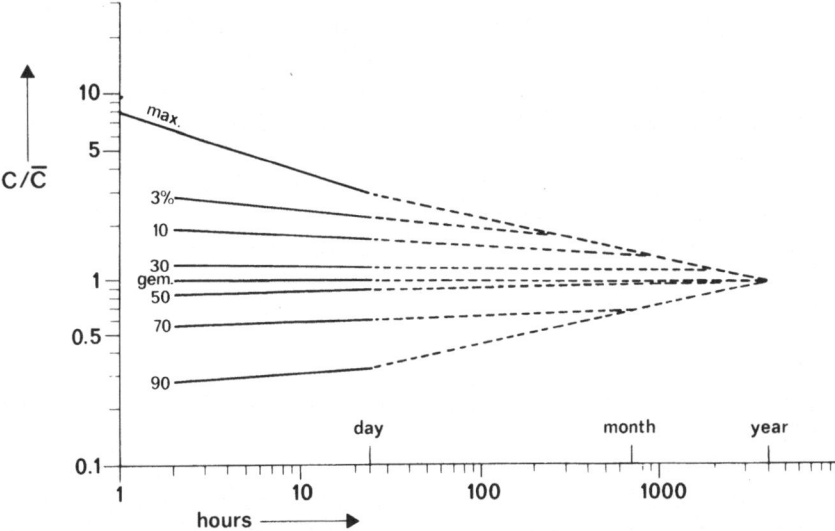

Fig. 11. The ratio of concentration and yearly average concentration $(c/\bar{c})$ as a function of the running averaging time. The curves pertain to different percentile values and are based on hourly SO_2 data from a monitoring station in an urban area in the Netherlands. The running average is indicated along the horizontal axis of the figure. It indicates for instance that the 3-percentile of the hourly average is approximately three times the yearly average concentration, and the daily (24 hr) average is $\sim 2.2\bar{c}$. The experience is that these graphs have some generality and might be used in similar locations, so that when only a yearly average is known, one is able to estimate percentile values for smaller averaging times. [Note that the mean value and the 50 percentile value do not coincide, indicating that the concentration distribution is (log-normally) skewed]. (Source: Modellen voor de berekening van verspreiding van luchtverontreiniging, Staatsuitgeverij, Den Haag 1976)

without much evidence that they led to significantly better results. From the point of view of policy making this was an unsatisfactory situation and a general need for verification, evaluation and standardisation of models existed. Both in the U.S. (St. Louis) [25] and in Europe (Frankfurt) [26] evaluation studies where carried out. In Frankfurt sixteen Gaussian plume models were tested and compared with each other and with observations of ground-based sulphur dioxide concentrations. The major conclusions of this study were that (i) most models reproduced the position of the concentration maxima quite well, (ii) the differences between predicted maxima of the models was of the same order as the differences between measured and predicted maxima, and (iii) the calculated city-centre to suburb gradient was in nearly all cases larger than the measured ones.

The study was carried out in 1976. Since then many institutions over the whole world have included urban air pollution models in their air quality management systems (AQMS). It is now generally agreed that – on a yearly basis – these models are able to predict ground level concentrations, which lie within a factor of two of the observed values.

Medium Range Transport Models

Here the assumption that meteorological conditions can be considered stationary during the dispersion process has to be abandoned. Transport times are of the order of a day so that the diurnal change in atmospheric stability (see section "Atmospheric Physics") should be included in the model. Also the horizontal transport wind varies both in time and over the modelling region. Because transport times are much longer than in the models of the previous section it appears that deposition and (photo) chemical processes cannot be ignored. Another complicating matter is that the terrain properties cannot be regarded homogeneous anymore. This and the fact that solutions can now only be obtained by solving partial differential equations, makes that modelling in this field is still not trivial and remain in the "research stage."

Medium range transport models have three major applications:

1. Combined with a model of atmospheric chemistry they are in principle able to predict (peak) photo-oxidant, or ozone levels during episodes of warm, sunny wheather, which may last a few days.

2. The high resolution (5–50 km) allows that these models can handle dispersion in topographically induced flows (mountainous area, coastal circulations).

3. Their detailed structure makes it possible to test model assumptions in other less complicated models, i.e., they may serve as a research tool. Moreover, some medium range transport models have such detailed vertical structure that it is possible to include sources at the approximately correct release height so that the influence on ground level concentrations and deposition from sources (or source categories) at various heights can be investigated.

The basic transport equation is (49), which neglecting along wind diffusion becomes

$$\frac{\partial C}{\partial t} + U\frac{\partial C}{\partial x} + V\frac{\partial C}{\partial y} + W\frac{\partial C}{\partial z} = \frac{\partial}{\partial y}\left(K_y \frac{\partial C}{\partial y}\right) + \frac{\partial}{\partial z}\left(K_z \frac{\partial C}{\partial z}\right) + S, \quad (88)$$

where S denotes the various sources and sinks. The three component of the mean wind velocity, (U, V, W), and the eddy diffusion coefficients are generally a function of space and time variables. An operational requirement is that these data can be easily and routinely obtained at any place. Therefore the major source of these data should be the standard weather data from the synoptic meterological network of the World Meteorological Organisation (W.M.O.). Of course global dynamic circulation models, such as provided by the European Centre for Medium Range Weather Forecasting (E.C.M.W.F.) could provide us with these data, but up to now the required spatial resolution is too crude and also data in the atmospheric boundary layer are still not that accurate to be used in atmospheric dispersion models.

Consequently procedures should be developed to derive turbulence and wind data etc. from routine meteorological observations.

Turbulence Data

In Fig. 12 a scheme is given, which roughly shows how to determine diffusion characteristics, both for urban and medium range models. A first step is to determine the intensity of the turbulence which is closely related to the atmospheric stability and which can be expressed in the Obukhov length, L. This parameter is required when we want to determine the height dependence of the vertical eddy diffusion coefficient, K_z (cf. Eq. (41)). (The horizontal diffusion coefficient is more difficult to determine. One usually assumes it proportional to K_z).

Another important parameter is the friction velocity u_*. Both L and u_* can be obtained from (36) and (31), by an iterative procedure provided that the surface heat flux $H \equiv \overline{w'\theta'_0}$, the roughness z_0 and the windspeed at a specified height is

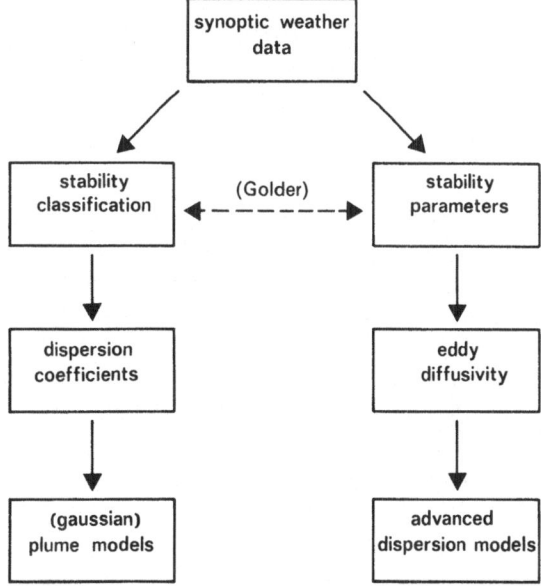

Fig. 12. Flow chart of the derivation of meteorological parameters for dispersion models

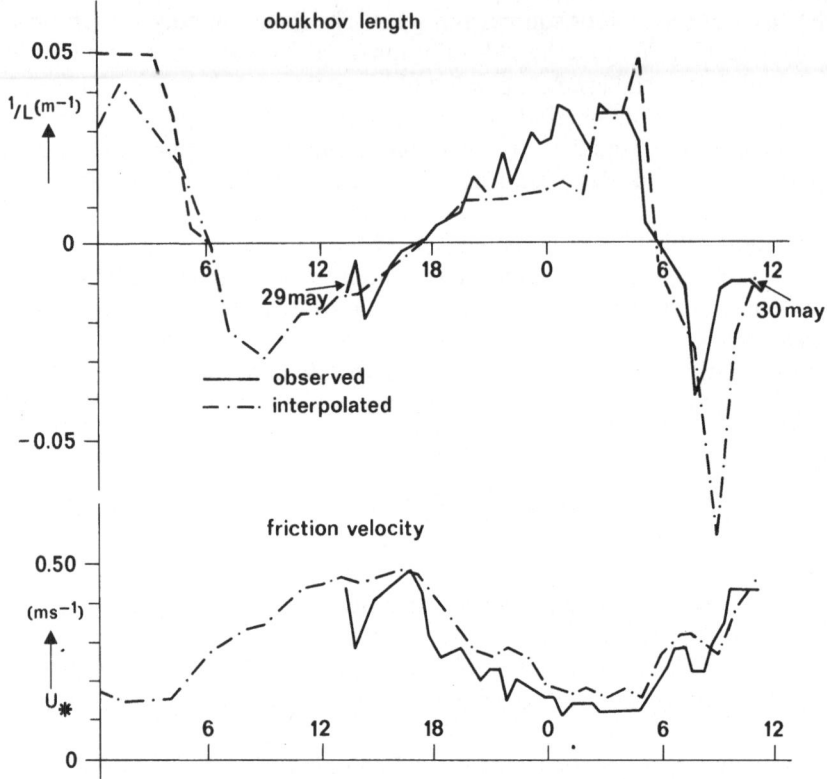

Fig. 13. Interpolated turbulence data (L and u_*) obtained from routine weather data (1978) are compared with direct observations (Data from Cabauw, KNMI, The Netherlands).

given. The last data can be obtained from routine weather data with hourly to three hourly intervals. An example of the diurnal variation in L and u_* as derived from routine data, is given in Fig. 13. They agree satisfactorily with in situ turbulence measurements.

The roughness may be estimated from an inspection of the terrain [27, 28] and the heatflux may be derived from the energy exchange at the surface, (Eq. 4). The energy exchange is roughly determined by the surface composition and -temperature, and the incoming radiation which depends on solar elevation and cloud cover. Procedures have been recently developed to estimate these [29]. An alternative is to determine first the Pasquill stability class from routine weather data and then use Golder's relations [34] to determine L. The former method, however, is a more direct one, and contrary to the latter, does not suffer from a slight ambiguity. The method works quite well as is shown in Fig. 13. Work on general procedures to derive meteorological parameters for dispersion models from routine weather data is in progress [32, 30].

Deposition Processes

Material is brought down to the earth by wet and dry deposition processes. Wet deposition is the general process by which air pollution reaches the surface with the precipitation. The amount of deposited material depends on the nature of the precipitation (showers, rain), precipitation rate and the chemical properties of the pollutant. Most removal from the air takes place within clouds (rain out) and a smaller fractions is removed by precipitation falling through the polluted subcloud layer (washout), thus removing pollution through absorption by rain droplets. In order to quantify these removal processes and chemical transformations which go along with them a detailed analysis is required of gas-phase and aqueous-phase physics and chemistry. Of these processes still very little is known and also data are not available to include these processes in present atmospheric dispersion models. A common practice is to assume simply that the wet removal rate (S_w) is proportional to the concentration,

$$S_w = -k_w C. \tag{89}$$

The constant, k_w, may be chosen proportional to the rate of precipitation, $k_w = \alpha P$ with α constant and P the precipitation rate.

Also the dry deposition of small particles and gases at the surface contains a large chain of physico-chemical processes where the nature of the pollutant and the type and structure of the soil play an important role. However, here the dry deposition velocity concept [31] can be introduced and applied with some success. The dry deposition velocity, v_d, is defined as the ratio of the downward flux ($\phi \equiv \overline{w'c'}$) and the concentration at a specified level (usually 1 m),

$$v_d = \phi/C. \tag{90}$$

A large amount of field data on dry deposition (Fig. 14) exist so that the downward flux can be estimated from the ambient concentration (at 1 m) via (90). When the concentration is given at another height its value at 1 m can be estimated by assuming that the concentration profile is similar to the temperature profile (37) in the surface layer, whence

$$C(z) = C_0 + \frac{0.74 c_*}{k} \{\ln(z/z_0) - \psi_2(z/L)\}. \tag{91}$$

From this relation and the identity (cf. (35))

$$c_* u_* = \overline{w'c'} \equiv \phi = v_d C(z = 1\,\text{m}) \tag{92}$$

the concentration at any desired level in the surface layer can be determined.

The Wind Field

The wind field in medium range models can be obtained from limited area numerical models or directly from observations (or from a mixture of both). The fine resolution of the mesoscale requires a detailed analysis of the wind field below say, 2000 m. Upper air winds may be estimated from radiosonde data, or the pressure gradient at the surface (see (21)). The density in space and time of the latter data is often much better than that of the radiosonde. The wind field within the mixed- or inversion layer may be estimated from surface wind observations and assuming surface-layer similarity in the lowest part of the layer (cf. (36)). Higher up in the inversion-layer the wind velocity can only be roughly guessed by an interpolation procedure between the surface wind and the geostrophic wind

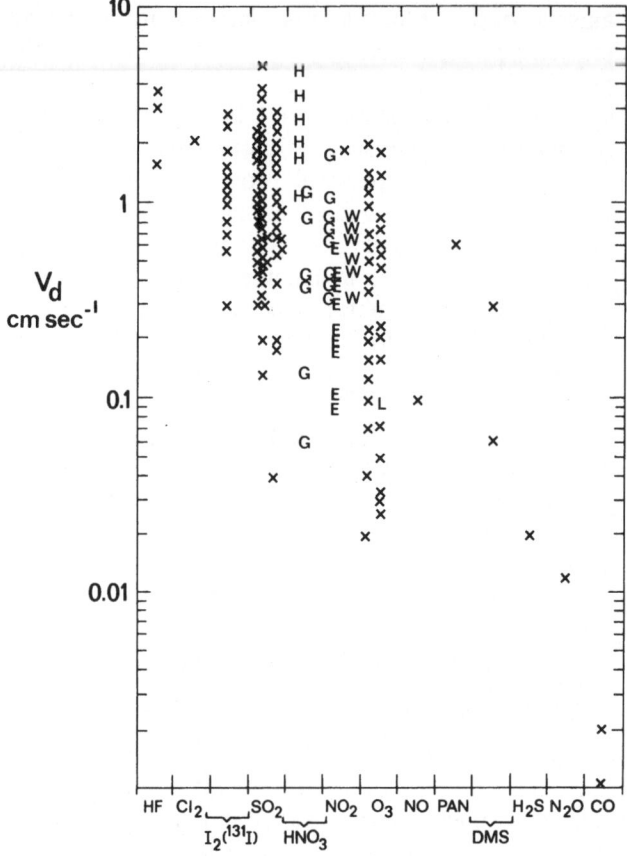

Fig. 14. Dry deposition velocity data for a number of pollutant ranked in terms of reactivity. (Source: ref. [40])

velocity [7], or calculated using statistical relationships between wind data at various levels [32].

It is clear that atmospheric modelling on this scale requires a combination of formidable knowledge of chemical and meteorological processes. Most transport modelling is still in progress and operational applications are sparse.

Long Range Transport Models

The dispersion of air pollution on the continental or global scale has been very intensively studied since 1965. Though at larger distances from the sources concentrations are too low to affect the environment directly by high concentrations in the air, the continuous and accumulative deposition of pollution may cause damage in sensitive areas. First reports of environmental deterioration came from the Scandinavian countries, which claimed that sulphuric acid deposition caused massive damages of their aqueous ecosystems. They claimed that the release of sulphur dioxide in the large European industrial areas contributed predominantly to acid deposition in Scandinavian lakes. Later, also other countries observed damage in lakes (Canada, United States), pine forests (West-Germany and Czecho-Slovakia) and materials (Acropolis (Greece) and historic monuments in many European cities) [33]. In order to obtain insight in the pathways of pollution, simple meteorological models were developed. These models were able to predict the deposition of the major pollutant (sulphur) over a period of one or two years, thus providing an insight in the contribution of one European country to the pollution in the other. These models have been extended recently, so that they also can cope with other pollutants and can include complex chemistry.

Simplifying factors in these models are that – as in the urban models – only statistical results are relevant, so that notwithstanding the lack of a precise knowledge of individual dispersion- and chemical processes one may hope that the averaging process smoothes some errors. Secondly, the transport time is of the order of 2–5 days, so that the effects of diurnal variation in stability and the vertical diffusion need not be considered in detail.

The modelling was carried out in the Lagrangian framework, i.e., a specific parcel of air was considered and its horizontal trajectory calculated. During its travel, uptake, deposition and chemical transformation was considered. Deposition and chemical transformation was initially described by linear decay. Exchange of contaminated air through the top of the parcel or side ways was neglected. Mathematically this can be formulated as

$$\frac{dC(\underset{\sim}{x}_0, t)}{dt} = -\left(\frac{v_d}{z_i} + k\right) C + \frac{Q}{z_i}, \tag{93}$$

where v_d is the deposition velocity, k the parameter which quantifies the loss due to wet deposition and chemical transformation. The source strength is denoted by Q and the initial particle position by $\underset{\sim}{x}_0 \equiv (x_0, y_0)$. The values of those parameters should be evaluated along the trajectory, which can be evaluated from

$$\underset{\sim}{X}(t) = \underset{\sim}{x}_0 + \int_0^t \underset{\sim}{u}_E(\underset{\sim}{x} = \underset{\sim}{X}(t'), t') dt', \tag{94}$$

where $\underset{\sim}{u}_E$ is the Eulerian wind field. The wind field is obtained from routine radio sonde ascents. Because the wind velocity changes with height there is some ambiguity in how a representative velocity for an air parcel extending to an altitude of ca. 1000 m should be determined. A much used procedure is to adapt the 850 mbar wind velocity by backing it over 20° and reducing its speed by 10%, so that it is thought to be representative for the transport of the air mass [35]. Computer models which analyses the radiosonde data can often provide wind data at intermediate pressure levels. The wind field at the 925 mbar level is sometimes used as a representative wind speed in trajectory models.

Some Lagrangian models include horizontal diffusion [36], by considering travelling puffs with (horizontally) Gaussian shaped concentration distributions, which gradually extend with travel time. Also trajectory models exist which account for the diurnal variation of the atmospheric boundary layer depth [37]. However, these additions can make the trajectory model formulation quite complicated, so that it looses its major attractivity: its simplicity.

We return to the model in its simplest form (93, 94) and extend the model to include a chemical transformation process (e.g., the oxidation from sulphur dioxide to sulphate) by means of a linear decay term. Because the two chemical components behave differently with respect to dry and wet deposition it is necessary to calculate the concentration budget separately. Denoting the two

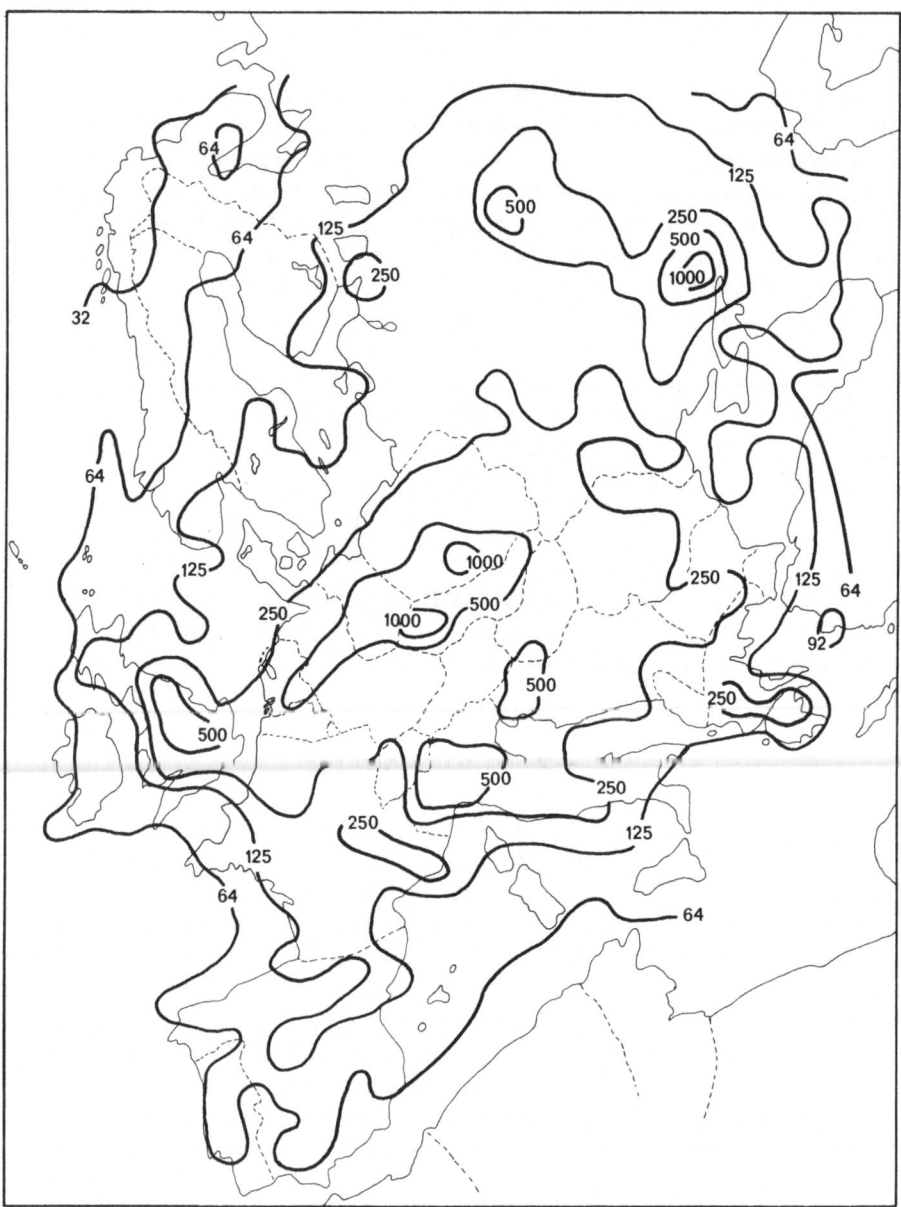

Fig. 15. Calculated average monthly dry plus wet deposition of sulphur from all emissions, in mg/m² as S. (Source: EMEP Report 1/82, NMI, Oslo)

components by C_1 and C_2, respectively, we have

$$\frac{dC_1}{dt} = -\left(\frac{v_d}{z_i} + k_1 + k_t\right)C_1 + \frac{Q_1}{z_i} \quad \text{and}$$

(95)

$$\frac{dC_2}{dt} = -k_2 C_2 + k_t C_1 + \frac{Q_2}{z_i}.$$

The indices 1, 2 refer to the two components and the parameter k_t "summarises" the complex of chemical reactions which transforms C_1 in C_2. Equations (94, 95) were used in the OECD programme on Long Range Transport of Air Pollutants (LRTAP) [38] to determine the deposition of Sulphur in Europe. Figure 15 depicts the total amount of deposited sulphur for the period October 1978–September 1980. Notwithstanding the simplicity of the model, its outcome agrees fairly with observed data of concentration and wet deposition (Fig. 16).

A convenient feature of the long term statistical trajectory model is that it enables one to estimate in any receptor location the relative contribution of all the sources in the area, so that the model is suitable for all kinds of applications in the field of control strategy and policy making. This is illustrated by taking the countries of Europe as emission and receptor locations and evaluating the deposition in one country as a result of the emissions in surrounding countries. From this we may compose a matrix which quantifies the relative contributions to the deposition in each country (Table 6).

Because in this simple model the source receptor relationship is linear (i.e., when an emission is reduced by say 30%, its contribution at any receptor location

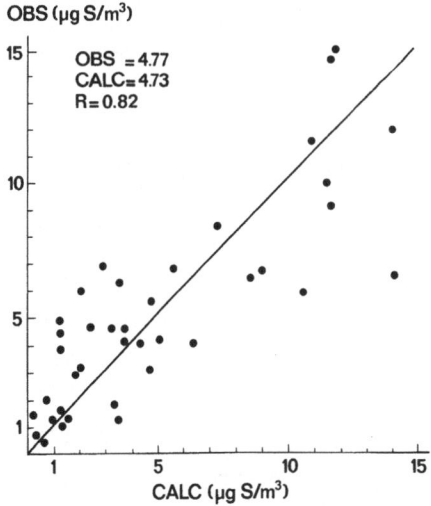

Fig. 16. Observed versus calculated two-year average concentrations of sulphur dioxide. The line represents perfect relationship. The correlation coefficient is denoted by R. (Source: EMEP Report 1/82, NMI, Oslo)

Table 6. National source-receptor matrix for Europe. The matrix is based on calculations with a Lagrangian transport model [38] for a two-year period (Oct. '78–Sept. '80). The countries are indicated by the official automobile registration abbreviation (except for the UK). The members in the table indicate the average monthly amount of total Sulphur deposition (unit: 100 metric tonnes as Sulphur). The average monthly emission per country is given in the column Q. The table should be read as follows: for example in the FRG ((D) in the 10-th row) 118 units per month is deposited which originates from the DDR ((DDR) in 9-th column). (Source: EMEP [38])

Receiver Countries × Emitter countries

	AL	A	B	BC	CS	DK	SF	F	DDR	D	GR	H	IS	IRL	I	L	NL	N	PL	P	R	E	S	CH	TR	SU	UK	YU	Q
AL	10	0	0	3	0	0	0	0	0	0	2	2	0	0	8	0	0	0	0	0	2	0	0	0	0	0	0	13	42
A	0	52	3	0	35	0	0	20	24	32	0	14	0	0	56	0	2	0	13	0	0	2	2	3	0	0	8	35	179
B	0	3	67	0	0	0	0	28	2	24	0	0	0	0	0	0	4	0	13	0	0	0	0	0	0	0	18	0	337
BC	0	0	0	153	8	0	0	2	6	4	7	12	0	0	8	0	0	0	7	0	34	0	0	0	3	12	2	45	417
CS	22	22	10	2	483	2	0	45	195	108	0	68	0	0	39	0	0	0	95	0	14	4	4	2	0	14	32	53	1250
DK	0	0	0	0	3	39	0	3	12	11	0	0	0	0	0	0	0	0	4	0	0	0	14	0	0	0	11	0	190
SF	0	0	2	0	8	0	77	5	19	15	0	3	0	0	2	0	2	2	15	0	2	0	0	7	0	43	14	3	225
F	0	0	33	0	7	0	0	629	20	98	0	0	0	2	41	4	12	0	4	0	0	66	0	0	0	0	99	6	1500
DDR	0	0	8	0	56	3	0	20	497	85	0	4	0	0	3	0	6	0	23	0	0	0	6	6	0	3	22	6	1667
D	3	8	37	0	48	4	0	108	118	561	0	6	0	0	24	5	23	0	18	0	0	6	2	6	0	3	72	15	1513
GR	3	0	0	34	4	0	0	3	3	3	93	5	0	0	14	0	0	0	2	0	8	2	0	0	4	4	3	27	293
H	0	9	0	0	46	0	0	7	18	13	0	194	0	0	27	0	0	0	21	0	16	0	0	0	0	4	3	75	625
IS	0	0	0	0	0	0	0	0	0	0	0	0	0	0	0	0	0	0	0	0	0	0	0	0	0	0	0	0	5
IRL	0	0	0	0	0	0	0	2	0	0	0	0	0	18	0	0	0	0	0	0	0	0	0	7	0	0	12	0	73
I	0	6	2	0	12	0	0	53	11	22	0	11	0	0	793	3	0	0	6	0	3	0	0	0	0	2	9	65	1833
L	0	0	0	0	0	0	0	3	0	2	0	0	0	0	0	3	0	0	0	0	0	0	0	0	0	0	0	0	20
NL	0	0	16	0	2	0	0	15	6	45	0	0	0	0	0	0	40	0	0	0	0	0	10	0	0	8	27	0	200
N	0	0	4	0	0	0	0	11	22	20	0	0	0	0	0	0	3	20	10	0	0	0	4	0	0	34	40	2	63
PL	0	7	7	3	136	8	2	26	213	30	0	43	0	0	18	0	7	0	565	0	13	2	0	0	0	0	32	40	1792
P	0	0	0	0	0	0	0	7	0	0	0	0	0	0	0	0	0	0	0	20	0	17	0	0	0	0	0	0	70
R	0	3	0	30	37	0	0	7	26	16	0	71	0	0	27	0	0	0	38	0	287	0	0	0	4	48	6	115	83
E	0	0	2	0	0	0	0	38	4	20	0	0	0	0	2	0	3	0	0	9	0	367	0	0	0	0	15	0	833
S	0	0	5	0	18	16	10	15	42	35	0	5	0	0	4	0	5	8	31	0	3	0	83	0	0	24	35	7	229
CH	0	0	0	0	2	0	0	23	3	13	0	0	0	0	47	0	0	0	0	0	0	0	0	14	0	0	5	3	48
TR	0	0	0	28	5	0	0	3	6	5	20	7	0	0	13	0	0	0	5	0	12	0	0	0	175	17	0	22	417
SU	3	14	18	69	189	25	50	67	283	190	22	159	0	2	94	3	23	5	386	0	205	10	38	3	453	610	103	208	6750
UK	0	0	7	0	4	0	0	27	11	21	0	0	0	7	2	0	5	0	2	0	0	4	0	0	0	0	675	0	2133
YU	3	14	2	32	38	0	0	20	25	22	6	72	0	0	131	0	2	0	21	0	26	7	0	0	0	8	7	557	1229

reduces also by the same amount), the effects of different emission scenarios can be directly quantified with respect to deposition.

An obvious extension is to consider other components as well. Some of them (ozone, ammonium, and nitrates) are also highly suspect in causing damage to forests and crops. Then, however, a more complex photochemical submodel is required which describes in more detail the successive reactions which lead to the formation of secondary chemical compounds such as nitrates and other organic nitrate compounds via gas phase chemistry of reactive hydrocarbons with nitric and sulphur oxides and gas phase aerosol interactions. As a consequence source-receptor relationships become non-linear so that for each source configuration separate calculations should be made.

Though trajectory models are a simple and powerful tool in the study of the transport of air pollutants, larger computer capacity facilitated the gradual introduction of long range Eulerian transport models, including complex chemistry. Both in Europe [39] and in the United States acid rain studies are undertaken and projects formulated where the development of these models is one of the major topics [40–42]. It is expected that these models become operational in the second half of the eighties.

Acknowledgement

The author summarises in this chapter major results in the field of air pollution research in the last two decades. His opinions and formulations are based on continuous and on-going exchange of information with many colleagues in his own institute and abroad.

References

1. Pasquill, F., Smith, F.B.: Atmospheric Diffusion, Ellis Horwood Ltd. 1983
2. Batchelor, G.K.: An Introduction to Fluid Dynamics, Cambridge Univ. Press 1970
3. Haltiner, G.J., Martin, F.L.: Dynamical and Physical Meteorology, McGraw Hill 1957
4. Nieuwstadt, F.T.M., van Dop, H.: Atmospheric Turbulence Modelling, Reidel 1982
5. Businger, J.A., Wyngaard, J.C., Izumi, Y., Bradley, E.F.: J. Atmos. Sci. *18*, 181–189 (1971)
6. Tennekes, H.: J. Atmos. Sci. *30*, 558–567 (1973)
7. Van Dop, H., de Haan, B.J., Engeldal, C.: Sci. Rep. 82–6, KNMI, De Bilt 1982
8. Nieuwstadt, F.T.M.: The nocturnal boundary layer, theory and experiments (thesis) KNMI, De Bilt 1981
9. Sutton, W.G.L.: Proc. Soc. A *182*, 48 (1943)
10. Nieuwstadt, F.T.M., de Haan, B.J.: Atmos. Environ. *15*, 845–851 (1981)
11. Monin, A.S., Yaglom, A.M.: Statistical Fluid Mechanics I, MIT 1973
12. Taylor, G.I.: Proc. London Math. Soc. *20*, 196–211 (1921)
13. Wax, N.: Selected Papers on Noise and Stochastic Processes, Dover 1954
14. Durbin, P.A.: NASA Rep. 1103, Lewis Research Center, Cleveland 1983
15. Janicke, L.: Proc. 12th ITM Nato-CCMS, Plenum Press 1982
16. Thomson, D.J.: Quart. J. Roy. Met. Soc. *110,* 1107 (1984)
17. Van Dop, H., Nieuwstadt, F.T.M., Hunt, J.C.R.: Phys. of Fluids **28** (6), 1639–1653
18. Richtmyer, R.D., Morton, K.D.: Difference Methods for Critical Value Problems, Interscience 1967

19. Prahm, L.P., Christensen, O.: J. Appl. Meteor. *16*, 896–910 (1977)
20. Orszag, S.A.: J. Fluid Mech. *49*, 76–112 (1971)
21. Gear, C.W.. Numerical Initial Value Problems in Ordinary Differential Equations, Prentice-Hall 1971
22. Parameters in the long-term Gaussian plume model: new recommendations, Institute for Applied Research (TNO) Delft 1984
23. Briggs, G.A.: Plume Rise, US Atomic Energy Commission Div. Tech. Inf. 1972
24. Larsen, R.I.: Rep. No. AP-89 US Environmental Protection Agency (EPA) Research Triangle Park NC 1971
25. EPRI, Overview, Results and Conclusions for the EPRI Plume model Validation and Development Project: Plains Site. EPRI-EA-30.74, RP1616-1 (1983)
26. Weber, E.: Air Pollution, Assessment, Meteorology and Modeling, Plenum 1982
27. Smith, F.B., Carson, D.J.: Boundary-Layer Met. *12*, 307–330
28. Van Dop, H.: Atmos. Environ. *17*, 1099–1105 (1983)
29. Holtslag, A.A.M., van Ulden, A.P.: Journ. Clim. Appl. Met. *22*, 517–529 (1983)
30. Holtslag, A.A.M.: Boundary-Layer Meteor. *29*, 225–250 (1984)
31. Thom, A.S.: Quart. J. Roy. Met. Soc. *98*, 124–134 (1972)
32. Irwin, J.S.: Gryning, S.E., Holtslag, A.A.M., Sivertsen, B.: Atmospheric Dispersion Modelling based on boundary-lay parameterisation, EPA report (to appear in 1985)
33. Acid Rain, A review of the phenomenon in the EEC and Europe. Graham and Trotman 1983
34. Golder, D.: Boundary-Layer Met. *3*, 47–58 (1972)
35. Pack, D.H., Ferber, G.J., Heffter, J.L., Telegadas, K., Angell, J.K., Hocker, W.H., Machta, L.: Atmos. Environ. *12*, 425 (1978)
36. Johnson, W.B., Wolf, D.E., Mancuso, R.L.: Atmos. Environ. *12*, 511–527 (1978)
37. Apsimon, H.M., Goddard, A.J.H., Wrigley, J.: Procs. of a CEC Seminar on "Radioactive releases and their Dispersion in the Atmosphere following a hypothetical reactor accident." Held at Risø, Denmark, CEC, Luxembourg 1980
38. Eliassen, A.: Atmos. Environ. *12*, 479 (1978)
39. Phoxa project. Programme of joint projects sponsored by the Umwelt Bundes Amt (UBA) (Berlin) and the ministry VROM (Den Haag) 1983
40. Regional Acid Deposition: Models and Physical Processes. NCAR tech. note NCAR/TN-214 (1983)
41. Environmental Research and Technology (ERT), Models for Long Range and Mesoscale Transport and Deposition of Atmospheric Pollutants Report SYMAP-101, Ontario Ministry of the Environment, Toronto 1982
42. Lamb, R.G.: A regional Scale (1000 km) Model of Photochemical Air Pollution. Part 1: Theoretical Formulation. EPA-600/3-83-035, Environmental Protection Agency. Research Triangle Park (NC) 1982
43. Gadd, A.J.: Quart. J. Roy. Met. Soc. *104*, 583–594 (1980)
44. Praagman, N.: Numerical solution of the shallow water equations by a finite element method. Ph. D. Thesis, Technological University of Delft, The Netherlands 1979
45. Fox, D.G., Orszag, S.H.: J. Comp. Phys. *11*, 612–619 (1973)
46. Chock, D.P., Dunker, A.M.: Atmos. Environ. *17*, 11–24 (1983)
47. Pepper, D.W., Kern, C.D., Long, P.E.: Atmos. Environ. *13*, 223–237 (1979)
48. Chock, D.P.: to be published in Atm. Environment (1985)

Notations

Some symbols have more than one meaning. Their significance will always be clear by the context in which they are used.

A   albedo

B   Bowen ratio

c   speed of light, concentration

$c_{p,v}$ specific constant of heat at constant pressure, volume
e turbulent kinetic energy
E latent heat of evaporation
f Coriolis parameter
g gravitational acceleration
G soil heat flux
h stack height, Planck's constant, Ekman boundary layer height
H sensible heat flux, release height in Gaussian plume model ($= h + \Delta h$)
k von Karman constant, Boltzmann constant
k_w decay parameter for wet deposition
$K_{m,h}$ eddy viscosity for momentum, heat
L Obukhov length
$m_{a,w}$ molecular weight of air, water
N Brunt-Vaisala frequency
u_* friction velocity
p pressure
P precipitation rate
Q source strength
Q^* net radiation flux
R universal gas constant
Ri Richardson number
R_L Lagrangian velocity correlation function
S solar constant, entropy
T absolute temperature
T_L Lagrangian time scale
t_k the Kolmogorov time scale
u,v,w components of wind velocity
z_i inversion layer height
z_0 roughness length
$\delta(x)$ Dirac delta function
Δh plume rise
ε dissipation rate, ratio of molecular weight of water vapour and air (m_v/m_a)
θ potential temperature
θ_* temperature scale in the surface layer
λ wave length, latent heat of evaporation
$\Gamma_{d,w}$ dry, wet adiabatic lapse rate
ν viscosity
ϱ density of air
$\sigma_{x,y,z}$ dispersion coefficients in the three coordinate directions
σ_w^2 Eulerian velocity variance ($\equiv \overline{w^2}$)
$\phi_{m,h}$ stability function in the atmospheric surface layer

The Mathematical Characterization
of Precipitation Scavenging and Precipitation Chemistry

J. M. Hales

Battelle
Pacific Northwest Laboratories, Richland, WA 99352, USA

Summary

This chapter examines the physical basis for precipitation scavenging, and provides an intermediate-level primer for scavenging calculations. The discussion is presented in the context of spatial and temporal variability of precipitation chemistry. It begins with a qualitative overview of behavior observed on precipitation-chemistry networks, and then provides a physically based rationale for these observations.

The majority of the text is addressed to examining how scavenging models are formulated using formal balances of mass, energy, and momentum in the atmosphere. Basic equations are presented in a highly visualized context, and procedures for "grooming" these basic forms to yield specific scavenging models are discussed. Following this, a number of illustrative examples are presented, and a general process for model selection is described.

Concluding portions of the chapter return to the initial theme of spatial and temporal variability, and give comparisons of selected model simulations with observed behavior. These comparisons are applied to evaluate uncertainties associated with present-day models, and specific areas for future improvement.

Introduction

The term *precipitation chemistry* pertains to that specific subcategory of atmospheric chemistry which deals with the chemical composition of condensed atmospheric water impacting on the Earth's surface. The chemical makeup of precipitation is a direct result of a natural air-cleaning process, often referred to as *precipitation scavenging*. The scavenging phenomenon is by nature highly interactive with a variety of atmospheric-chemistry components, and the associated physical and chemical interchange processes typically maintain the air-precipitation system in a highly dynamic state. This leads to significant variability of precipitation chemistry in both space and time – a subject which is an important underlying theme throughout this chapter.

Precipitation-chemistry attributes and their relationships to source properties and atmospheric variables are normally computed using *precipitation scavenging models*, which can be considered as multiphase extensions of their conventional air-quality counterparts. A major goal of this chapter is to provide the reader with a cross-examination of precipitation-scavenging models, to facilitate a broad understanding of the subject and provide a starting point for extended practical application. This text is also intended to provide a reference resource for a number of important "cornerstone" precipitation-scavenging equations, which can be referred to conveniently as needed by the user.

A number of initial comments are appropriate at this early point in the discussion. The first of these pertains to the frequently (but often rather vaguely) applied distinctions between "Eulerian" and "Lagrangian" models, and their "statistical" and "deterministic" applications. The term "Eulerian" pertains to those models whose governing equations are based on coordinate systems that are fixed with respect to the Earth's surface. This is often the most convenient way of addressing the modeled system, because it corresponds to a human viewer who is standing at a fixed point and "watching the atmosphere go by."

The term "Lagrangian" pertains to a mathematical coordinate system which moves in space with the average flow-velocity of the model's domain. Often this is a convenient reference, because it (sometimes) allows elimination of flow-related terms in the model's governing equations and provides a corresponding simplification. Because atmospheric wind-fields are so spatially variable, however, there are numerous possibilities for defining "average flow." Typically this problem is partially resolved by segregating the model system into individual plumes, subdividing further into individual plume segments or "puffs," and then defining appropriate localized mean velocities for each.

This artificial segregation necessitates, ultimately, a mathematical recombination of plumes to form composite concentration fields. This is usually performed by mathematically superimposing the individual components using simple addition – a practice which can result in rather substantial systematic errors if nonlinear relationships are involved. Moreover, the appropriate definition of a moving coordinate system is nontrivial, even for segregated systems. These two difficulties detract rather substantially from the Lagrangian approach, although the disadvantage is offset somewhat by relative computational simplicity in many cases. It is important for the reader to note in this context, however, that both Lagrangian and Eulerian models are based, in the limit, on the same fundamental equations and physical relationships. Moreover, a number of mixed models exist which simply utilize solutions to Eulerian-based equations and advect the derived values in a Lagrangian sense.

"Statistical" and "deterministic" models can be either Eulerian or Lagrangian in character, although a precise distinction between these two is by no means obvious. The term "deterministic" implies that the model is based on established physical cause-effect relationships, while "statistical" suggests that model elements are derived from stochastic properties. A model which utilizes a probability-density function to describe frequencies and durations of rainy periods, for example, normally would be placed in the "statistical" category.

There has been extensive debate (cf. Alp et al. [2]) regarding the degree of "statisticity" of models, and what constitutes a statistical model and what does not. It is important to note, however, that *all* models contain statistical components. The transport properties and reaction-rate coefficients typically employed within deterministic models, for example, are nothing more than descriptors of ensemble-average behavior of stochastic molecular processes.

In view of the above features it is not particularly useful to attempt to make any precise or dogmatic distinctions between these model classes. Because of descriptive convenience the following text will assume something of an "Eulerian-deterministic" tendency; the reader should bear the universiality of the mathematical approaches in mind, however, and note that the cited relationships will be generally useful to all types of modeling endeavors.

An attempt has been made to format this chapter so as to be readable at a variety of levels. Thus while (sometimes rather complex) mathematical expressions are used, this should by no means discourage use by the nonmathematical reader. This multilevel character of the chapter is facilitated by a highly visualized approach and a physically descriptive text surrounding the equations. As a consequence, nonmathematically oriented readers should expect to use the text to obtain a strong visual image of cause-effect relationships in precipitation chemistry. Their more mathematical associates can expect to use the chapter as a primer on the process of scavenging modeling. The experienced modeler should find the text useful as a reference source for key mathematical relationships.

The chapter begins by giving a qualitative overview of precipitation-chemistry observations, with particular reference to their variations in space and in time. This variability, which is attributed to mechanistic phenomena in the precipitation-scavenging process, is used as a lead-in to the scavenging-modeling portions of the text. The scavenging-modeling discussion is subdivided logically into a number of parts. The first of these is an overview of the fundamental governing equations describing material, energy, and momentum balances associated with the scavenging process. These equations are shown to contain internal expressions describing physicochemical microprocesses, such as pollutant pickup by individual raindrops and in-drop chemical reactions, and the succeeding section deals directly with these "microscopic" aspects.

The next section of the chapter demonstrates how the microscopic phenomena can be combined with the balance equations, and the composite equations can be solved to produce desired fields of precipitation-chemistry attributes. Within this discussion a formalized approach is emphasized, which implements a systematic "grooming" of general equations to meet specific demands. This is followed, in turn, by the description of a decision-chart technique for model selection.

As a final note it should be emphasized that several excellent discussions of precipitation scavenging exist, and that this text is intended to be complementary to these literature resources. In particular, a well-written discussion of gas-scavenging microprocesses has been presented by Peters [82], and a more meteorologically oriented treatment of general scavenging phenomena has been given by Slinn [102]. Moreover, there are several useful and comprehensive texts on cloud-physical processes, the book by Pruppacher and Klett [84] being a particularly apt example. Finally, the classic text, *Transport Phenomena*, by Bird et al. [9] provides an invaluable reference on material energy, and momentum balances, a subject that is of central importance to this work. The present text is hoped to be a useful addition to these references in the sense that it provides a systematic and balanced merger of several of these contributions on a somewhat less detailed basis. The reader is encouraged to pursue these resources while studying the material of this chapter.

Spatial and Temporal Variability: A Qualitative Overview

In addressing the question of precipitation chemistry's variability, it is important to note that all chemical constituents existing in rain and snow occur as the consequence of a complex chain of atmospheric material-exchange processes. These individual processes exhibit inherent fluctuations in space and in time, and

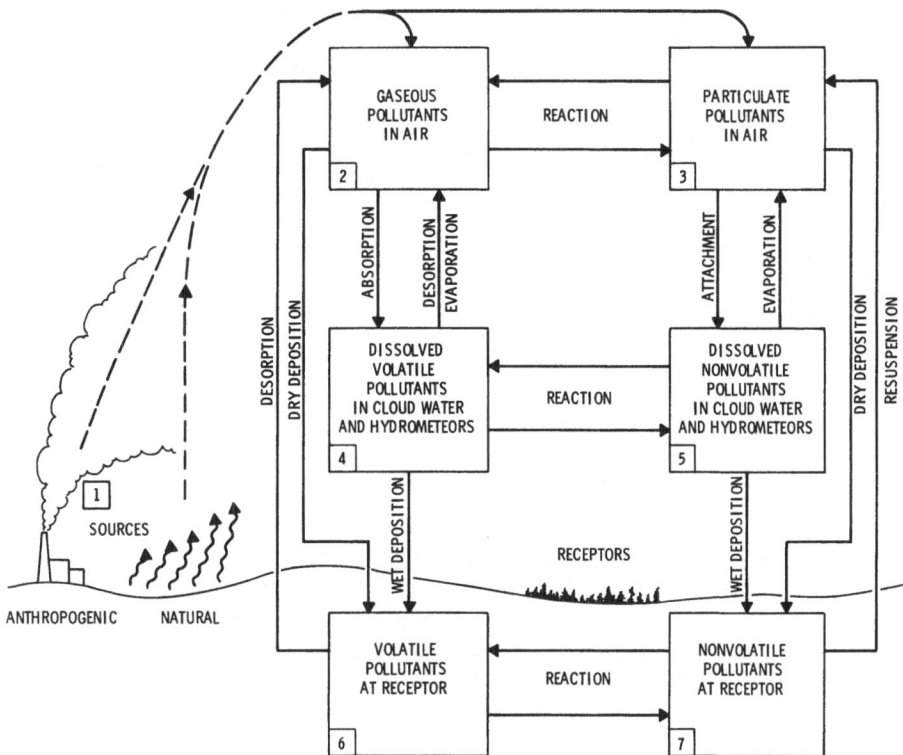

Fig. 1. Schematic of atmospheric source-receptor pathways

the variability observed in precipitation chemistry is a composite reflection of this sequence of events.

The complex of atmospheric interactions leading to wet-delivery of pollutants to the ground is shown schematically in Fig. 1, which depicts a given pollutant as it is emitted from a source and ultimately delivered to a receptor, via the atmosphere. Important points to note from this diagram are the *competing* effects of wet and dry deposition, and the potential for reversible *cycling* of pollutants through the atmosphere. It is important also to note that Fig. 1 depicts a general pollutant *material balance* in the atmosphere, and that *mathematical* models of pollution behavior can be formulated from this *conceptual* model, simply by setting forth equations characterizing chemical flows through each of the boxes. Much more will be said on this subject in the following section of this chapter; for the present it will suffice to note that both spatial and temporal variability of precipitation chemistry are related intrinsically to atmospheric material balances, and represent a composite reflection of the source-receptor experience.

From this qualitative discussion it is easy to list some of the more important contributors to precipitation chemistry's spatial and temporal variability:
- variability associated with source fluctuation and configuration,
- variability associated with normal atmospheric advection and mixing processes,

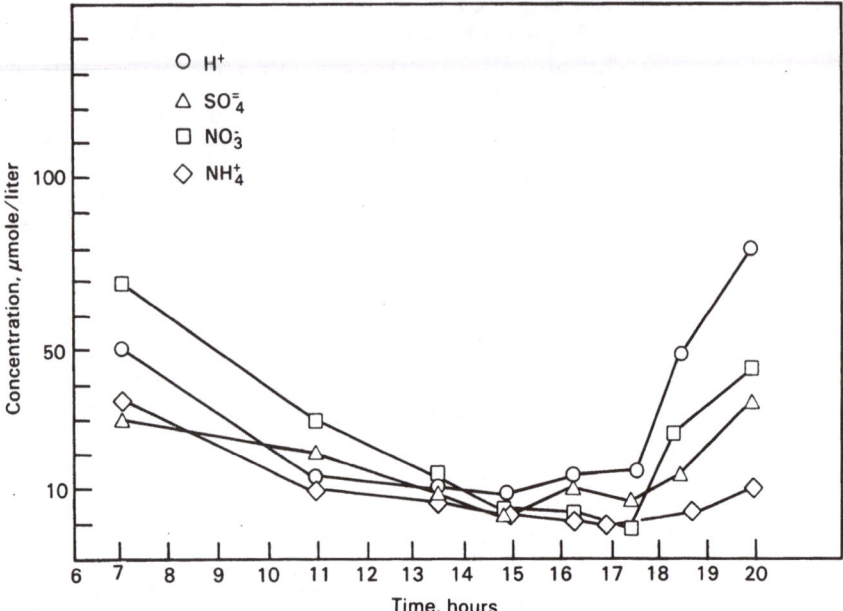

Fig. 2. Time-series of precipitation chemistry data for a single rain event. Data from Penn State MAP3S Station. (Adapted from de Pena et al. [19])

- variability induced by storm dynamics,
- variability caused by atmospheric transformation processes prior to the precipitation event,
- variability associated with microscale physicochemical processes in the cloud environment,
  and
- variability caused by pollutant depletion via wet- and dry-removal processes.

These features are difficult to isolate, and their relative effects will vary depending on the averaging times associated with the precipitation-chemistry measurements at hand. In performing and assessing scavenging calculations, however, it is important that one keeps these factors in mind and attempt to define the spatial and temporal averaging times appropriate to his own requirements.

In addressing the *temporal* nature of precipitation chemistry, it is usually observed that different components of the above list gain or diminish in relative importance as averaging times increase. This is demonstrated by the sequence of Fig. 2 through Fig. 6, which shows time-series of precipitation-chemistry data for progressively expanding "time windows" and averaging times. Figure 2, for example, shows the results of a sequential sampling of rain from a single precipitation event observed at the Pennsylvania State University MAP3S site [75, 19]. This is a rural sampling location in central Pennsylvania and is considered to be a representative indicator of regional behavior. Key features to note from this figure are the pronounced variability of concentration during storm passage, and the obvious continuity of the time-concentration curves. The exhibited behavior, involving high initial concentrations followed by reduced

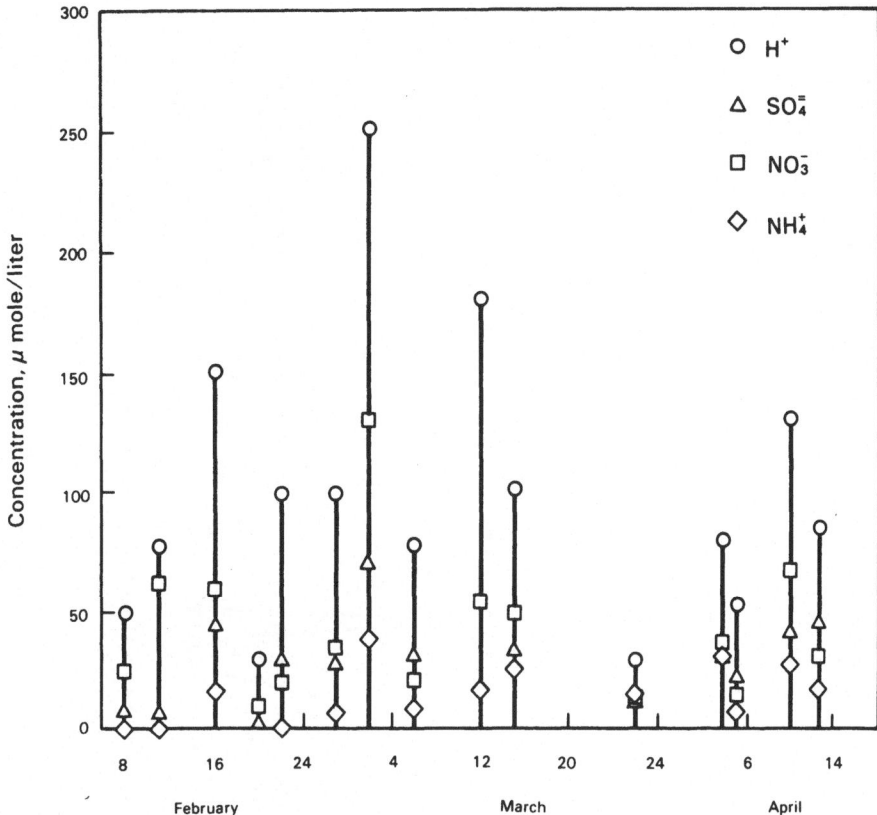

Fig. 3. Time-series of event-averaged precipitation chemistry data. Data from Penn State MAP3S Station

values and then recovery toward the end of the storm, is not uncommon in sequential observations; however, a large variety of time-concentration curve shapes has been noted in past studies.

Figure 3, also pertaining to the Pennsylvania MAP3S site, is a typical result obtained when one averages precipitation-borne pollutant concentrations over entire precipitation periods, and plots several events in sequence. Here discrete plotting is necessary, owing to the episodic nature of precipitation. In view of our previous discussion, the existence of large fluctuations – in spite of the longer averaging times – should not be surprising: the within-event variability removed by the event-averaging process has been supplanted by new components which are longer-term in nature and would not have been obvious if the time window had not been expanded.

Figure 4 pertains to an expanded data set which originated from this same sampling site, but now has been averaged over one-month periods. At this point the averaging process appears to have smoothed the concentration excursions somewhat, although new components of variability are indeed evident. The pronounced seasonal cycling of some of the species (hydrogen, sulfate, and ammonium ions) is typical of observations in northern and temperature latitudes,

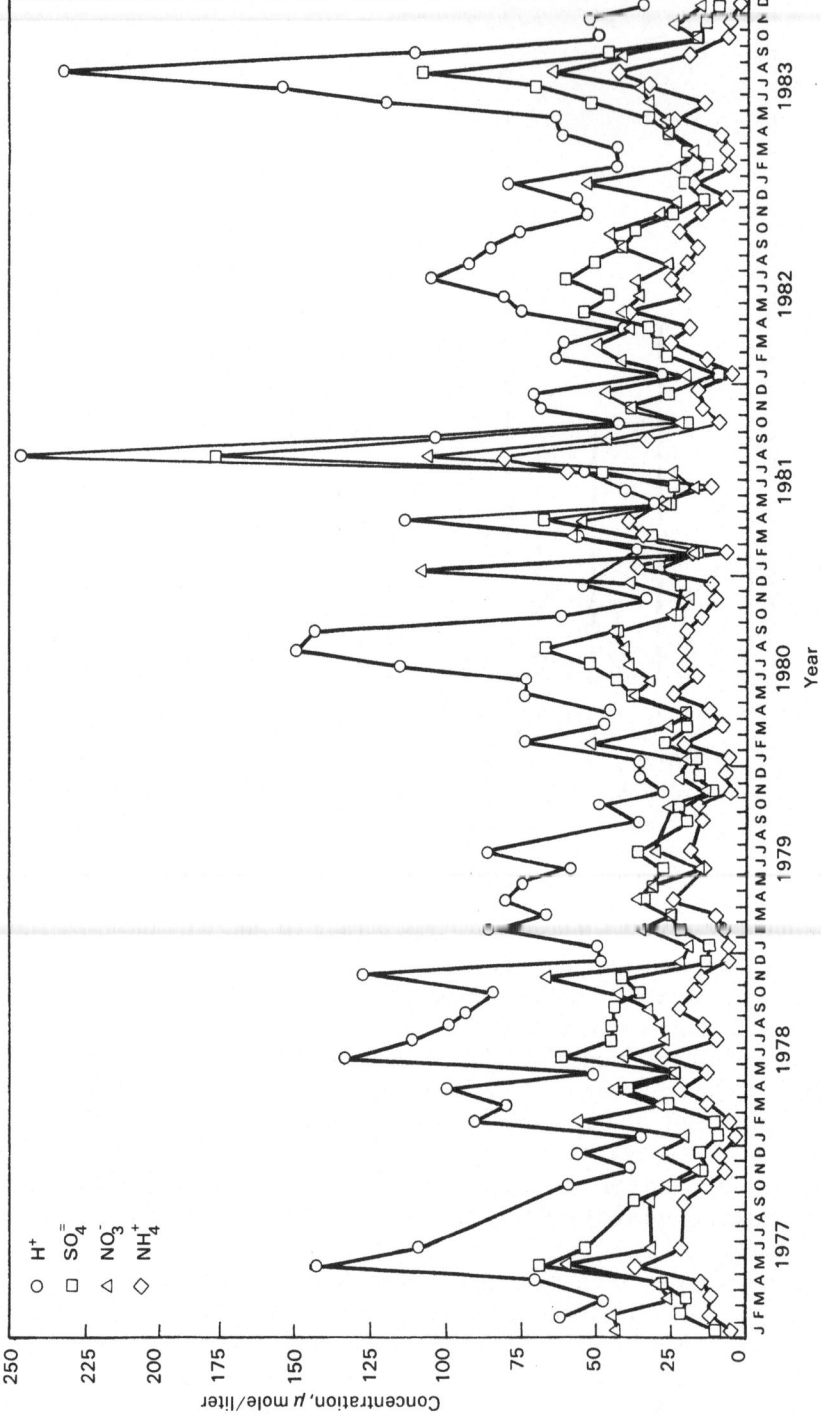

Fig. 4. Time series of monthly-averaged precipitation-chemistry data. Data from Penn State MAP3S Station

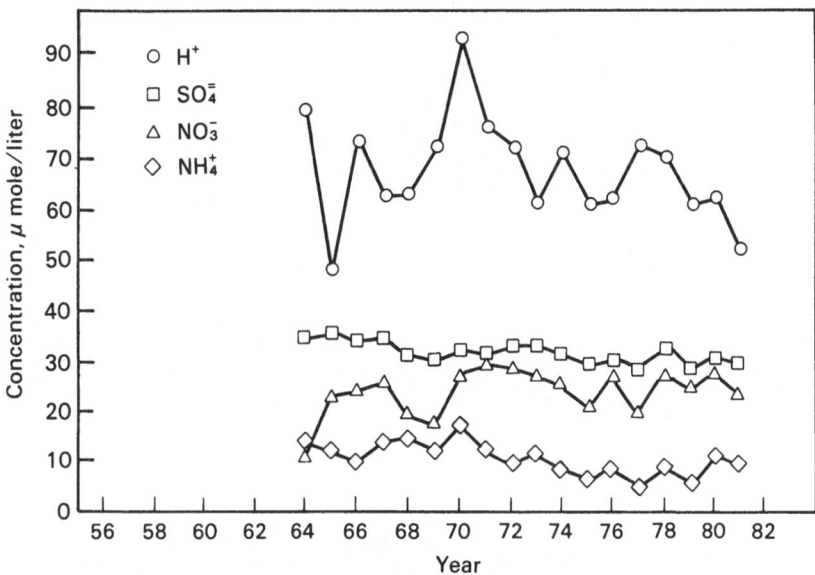

Fig. 5. Time series of annually averaged precipitation chemistry data. Data from Hubbard Brook, New Hampshire Station. (Adapted from Likens et al. [71])

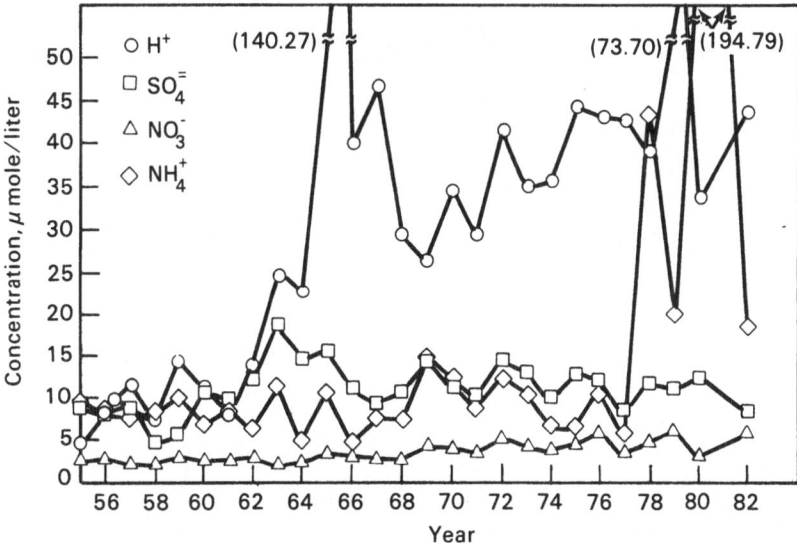

Fig. 6. Time series of annually averaged precipitation-chemistry data. Data from Forshult EACN Station in Central Sweden

although geographical differences do exist. Precipitation-borne ammonium-ion concentrations in northern-European samples, for example, appear to maximize during winter months, in direct contrast to North American behavior.

If one were to time-average Fig. 4's data over yearly periods much of the observed variability would be removed. This should be viewed with some caution,

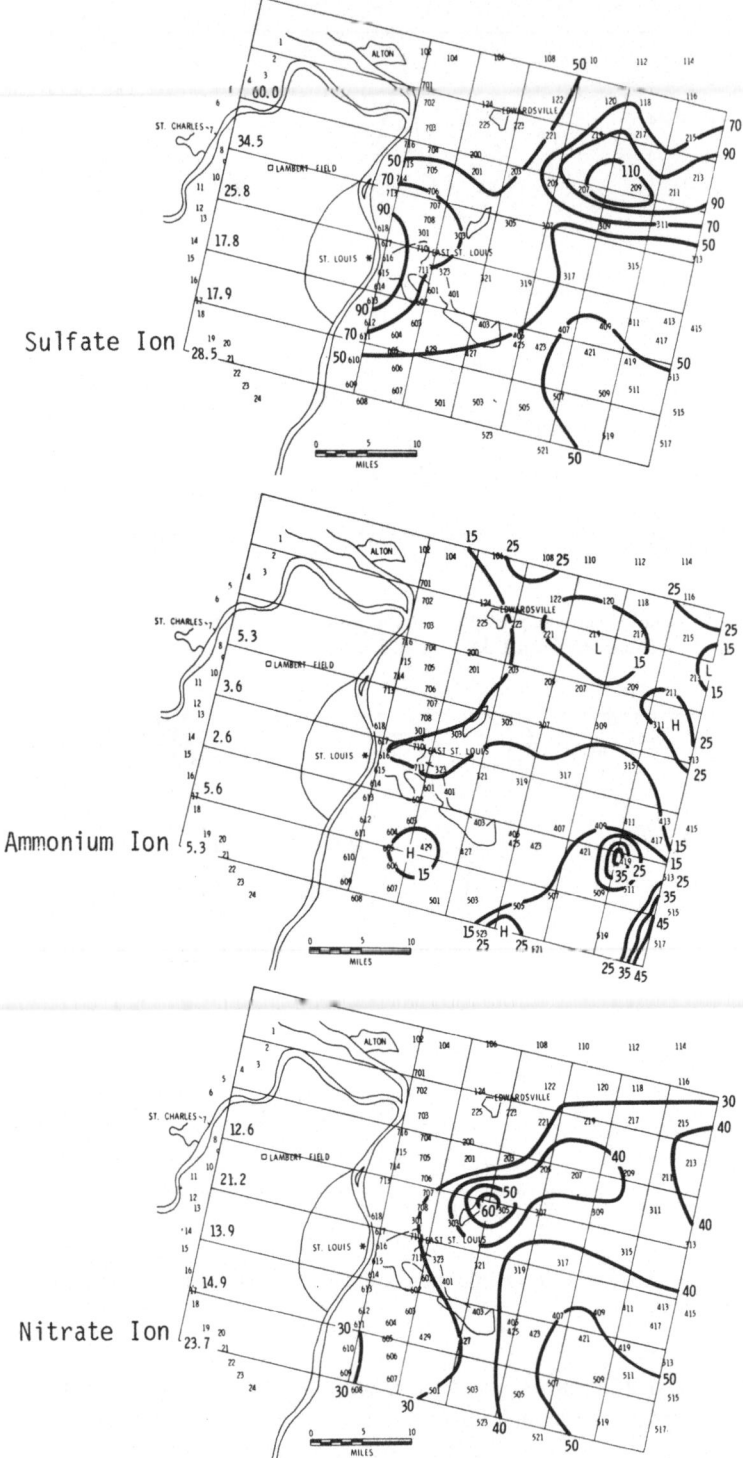

Fig. 7. Spatial distributions of precipitation chemistry delivered by a summertime convective storm near St. Louis, MO [45]. Concentrations in micromoles per liter

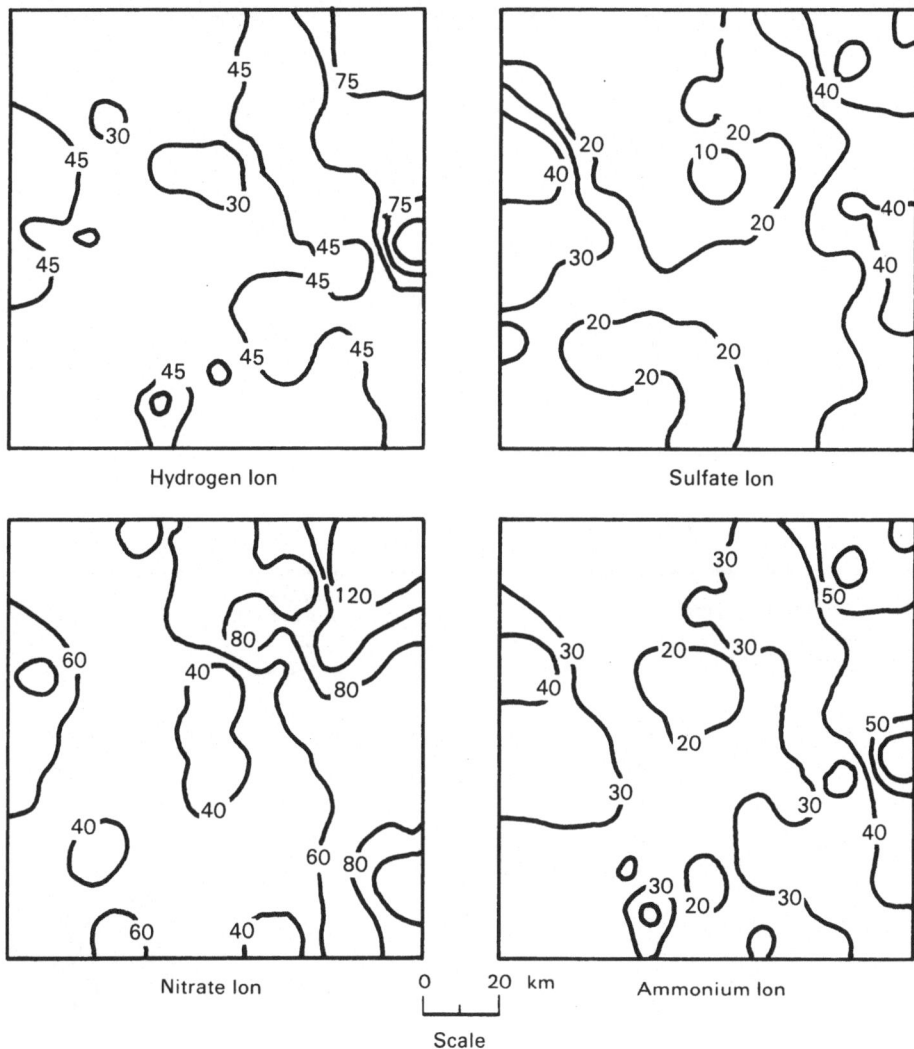

Fig. 8. Spatial distributions of precipitation chemistry delivered by an occluded springtime frontal storm in northeastern Indiana [15]. Concentrations in micromoles per liter

however, since longer-term data appear to indicate rather substantial excursions. Evaluation of these longer-term periods is hampered, however, by the fact that early collection and analysis techniques frequently gave rise to data of questionable quality, and it is often highly uncertain just how much variation was induced by measurement errors as opposed to actual atmospheric phenomena. An example of longer-term North American data is given in Fig. 5, which is a presentation of annually averaged results collected at the Hubbard Brook, New Hampshire site during the period 1964–1982 [71]. This plot possesses segments with rather pronounced upward and downward excursions, behavior which is typical of longer-term European data as well. Figure 6 presents a time-series plot of data obtained from the European Air Chemistry Network site at Forshult in central

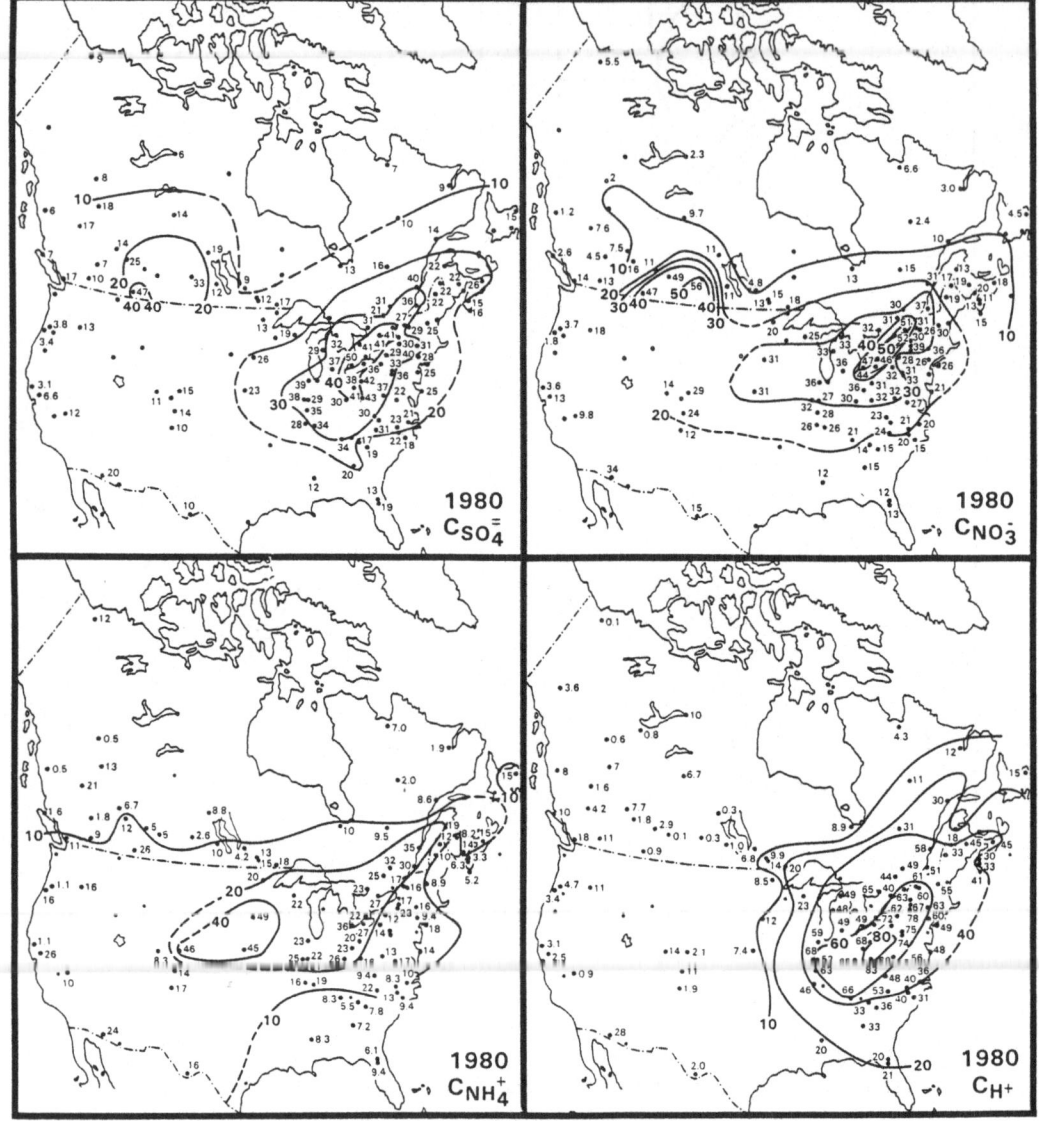

Fig. 9. Spatial distributions of annually averaged precipitation chemistry in North America during 1980 [5]. Concentrations in micromoles per liter

Sweden, which illustrates this point. Questionable features of data quality combined with the noted tendency for excursions severely complicates the evaluation of long-term deposition trends, and as a consequence major uncertainties exist in this area (e.g., Granat [38], Rodhe and Granat [86], Hansen and Hidy [50], Cogbill et al. [14], Lilljestrand and Morgan [72]).

In addressing *spatial* variability, it should be noted that point-to-point variations in rainborne pollutant concentrations will be strongly related to

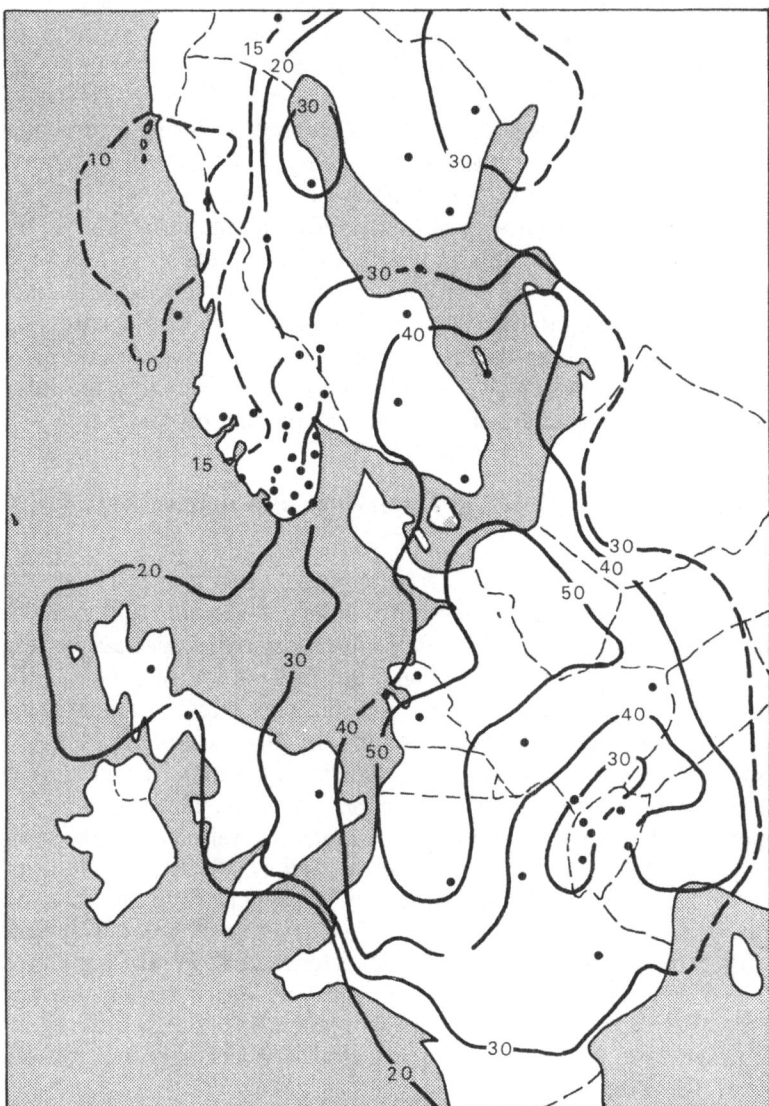

Fig. 10. Spatial distributions of annually averaged precipitation-borne sulfate ion in Europe during 1974 [79]. Concentration in micromoles per liter

temporal variability in most cases. Spatial patterns of short-term averaged precipitation chemistry, for example, are usually much more intricate than those for annual averages. Some feeling for this can be obtained by comparing the event plots of Figs. 7 and 8 with the annual averages shown in Figs. 9 and 10. The extended annual plots seem to be comparatively well-behaved in the sense that the data are sufficiently correlated spatially to permit the construction of isopleths from relatively sparse sampling points. Moreover, a relatively consistent picture typically emerges upon comparison of spatial patterns for multiple-year data sets.

This is usually not the case with event measurements. The general subject of variability and its relationship with spatial and temporal scales is discussed in an excellent introductory treatment by Finkelstein [32]. As noted by this author, our techniques for interpreting spatial variability are currently in an evolving stage, and much more sophisticated analyses can be expected during future years.

This brief overview of chemical variability has focused specifically on hydrogen, ammonium, nitrate, and sulfate ions. These inorganic nonmetallic species were chosen for illustrative purposes only, and the reader should observe that vast numbers of both organic and inorganic substances occur, and have been measured, in surface precipitation. In conclusion to this section it can be noted that our listing of sources of spatial and temporal variability seems adequate to explain many of the features of observed data, at least in a qualitative sense. In the section immediately following we shall examine procedures to quantify this behavior, using mathematical characterizations of the scavenging process.

Precipitation-Scavenging Modeling: Formulation of Governing Equations

Introduction

In the previous section it was mentioned briefly that *mathematical* models of precipitation scavenging can be formulated on the basis of *conceptual* models, such as the one shown schematically in Fig. 1. There is of course a wide latitude of procedures for accomplishing this task, and this has led to a rather large array of mathematical models available for precipitation-chemistry estimation.

Several important points can be observed in this context. The first of these is that there are numerous interties between individual mechanisms in the scavenging sequence, and that it is often difficult to isolate single effects from the composite whole. Wet chemistry, for example, affects and is affected by cloud-physics processes, which are in turn influenced by wind fields and thermodynamics. Sometimes referred to as the "scavenger's snarl," this complex of interactions usually forces calculations to be performed in terms of a "gestalt" concept, rather than as a collection of separable parts. Along these lines it can be observed that complete scavenging models, by necessity, contain most of the elements of clear-air pollution models in addition to their essential wet-chemistry and wet-removal terms.

Another important observation is that scavenging models can differ from one-another depending on which regions of Fig. 1 they are intended to simulate. A comprehensive model, for example, will simulate the input to the atmosphere from the sources, and carry the calculations to the pollutant's delivery at the receptor points. Another model, however, might utilize observed concentration data at the storm's inflow region, thus beginning the model's domain here, operating over a smaller portion of the conceptual region, and avoiding source simulation and the treatment of clear-air interactions. In general the uncertainty level of a model will increase with the extent of its conceptual domain, simply because of the cumulative errors associated with simulating additional pathways.

The general procedure for mathematical-model formulation is described schematically in Fig. 11. In box 2 of this figure, mathematical expressions derived

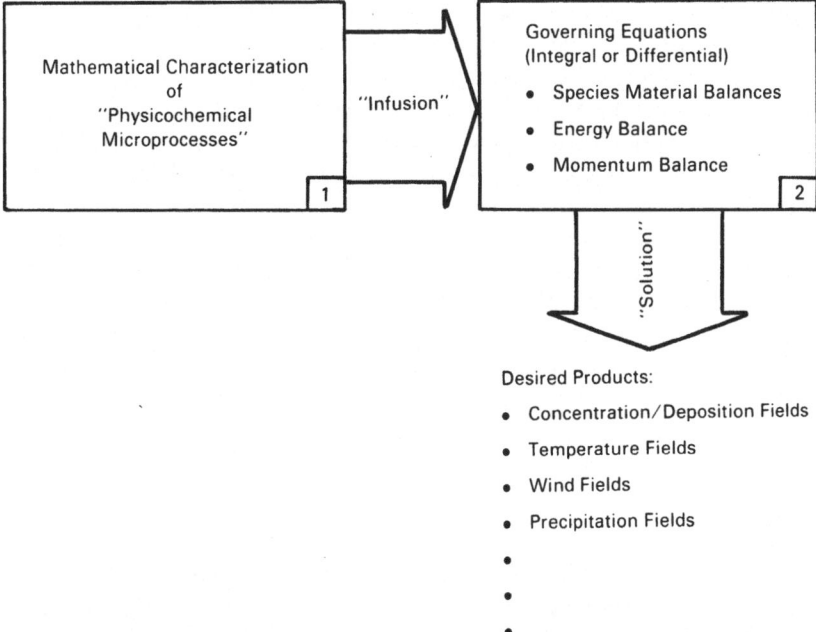

Fig. 11. Schematic of general scavenging-model development process

from physical-balance relationships are set forth for ultimate solution. As noted previously, these *mathematical* formulae are derived from *conceptual* models of physical behavior; the "species material balances" in box 2, for example, could be derived directly from Fig. 1 by summing inflows and outflows of pollutants on a balance sheet, although more fail-safe and scientifically viable procedures are recommended. Once formulated, these "governing equations" will invariably contain undefined terms associated with internal processes such as chemical reaction and aerosol capture. These "physicochemical microprocesses" must be characterized mathematically and inserted in the governing equations before solutions can be obtained. Upon final solution, subject to the appropriate initial and boundary conditions, the successful model can be expected to yield fields of the desired physical attributes, including those for precipitation chemistry.

The goals of the present section are to familiarize the reader with the general procedures for achieving the above ends, and to set forth in reasonably concise form a number of useful relationships for this purpose. In so-doing it is intended that this will be a valuable procedural outline for the inexperienced modeler, and will also be a useful equation reference for those who are more familiar with the field. The inexperienced reader may find some initial difficulty with the differential equations and associated vector notation employed in early portions of this discussion. He is advised to ignore this feeling of discomfort, because the text is written to give a strong visual image of the physical relationships associated with these terms. As noted in the *introduction*, it is intended that this chapter be readable on several levels of scientific detail, depending upon the desires of the user.

As indicated in Fig. 11, scavenging models can be formulated by performing balances of pollutant species, energy, and momentum, and such balances can be "differential" or "integral" in character. Differential and integral balances differ in the size of the elemental volume used for the operation. *Differential* balances are normally based on small volume increments and yield *differential equations*, which must be integrated subsequently to provide the desired computations of concentrations and removal rates. In this sense the model's total spatial domain can be considered to be "packed" by a large number of very small volume elements. *Integral* balances typically are performed over much larger regions, and result in *integral equations* or else algebraic forms derived from some sort of implied integration process. In integral material balances the element domain usually coincides with the model's spatial domain; thus only one volume element is considered. Quite often, however, property balances are mixed in nature and yield correspondingly mixed mathematical forms.

Material Balances

The preponderance of today's scavenging models are formulated from material balances only, and do not include direct calculations of energy or momentum. This is acceptable as long as external data are available to supply temperature and wind fields for model input. A general form of the differential material balance is given as follows:

$\dfrac{\partial c_{Am}}{\partial t} =$	Time rate of change of concentration of species A in medium m in a small volume element of atmosphere (moles/cm^3 s) =
$-\nabla \cdot (c_{Am}\underline{v}_{Am})$	$-$ rate of loss of species A in medium m by transport through walls of volume element (moles/cm^3 s)
$+w^*_{Am}$	$+$ rate of gain of species A in medium m by interphase transport of A to medium m from other media (moles/cm^3 s)
$+r^*_{Am};$	$+$ rate of gain of species A in medium m by chemical reaction within the element; (moles/cm^3 s)

A = 1, 2, ..., total number of species;
m = 1, 2, ..., total number of media.

By "medium" we mean air, rain, cloudwater, snow, ..., or whatever set of physical media are deemed necessary to simulate the situation at-hand; for example, if one wished to distinguish only between rainwater and air media, then the following equations could be written:

$$\frac{\partial c_{Aair}}{\partial t} = -\nabla \cdot (c_{Aair}\underline{v}_{Aair}) + w^*_{Aair} + r^*_{Aair}, \tag{2}$$

$$\frac{\partial c_{Arain}}{\partial t} = -\nabla \cdot (c_{Arain}\underline{v}_{Arain}) + w^*_{Arain} + r^*_{Arain}. \tag{3}$$

One should note that, in the simple rain-air system depicted by these equations, the interphase transport of pollutant from the air to the rain must match that from the rain to the air; that is

$$w^*_{Aair} = -w^*_{Arain}.$$ (4)

It is important to recognize in this context that material balances can be formulated, not only for *pollutant species*, but for the *media* as well. Indeed, some classes of scavenging models generate cloud and precipitation fields as part of their computed output, and do so by solving material-balance equations for clouds, snow, and rain. One particularly important medium equation in this respect is the material balance for *air*, which can be derived directly from the more general forms above.

$$\frac{\partial c_{air}}{\partial t} = -\nabla \cdot (c_{air} \underline{v}_{air}).$$ (5)

The velocity vector $\underline{v}_{Am}$ in Equation (1) can be subdivided into its x-, y-, and z-components:

$$\underline{v}_{Am} = U_{Am}\underline{i} + V_{Am}\underline{j} + W_{Am}\underline{k},$$ (6)

where the component velocities can vary markedly from one medium to the next. Usually U_{Am} and V_{Am} are interpreted as normal wind velocity components plus diffusional components, while W_{Am} must contain components for precipitation fall-speed whenever it corresponds to hydrometeors such as rain or snow. It should be noted also that multiple phases and the interphase-transport terms are key ingredients distinguishing this type of scavenging model from clear-air pollution models. This constitutes, essentially, the addition of a set of "wet" species-balance equations to their "dry" counterparts.

It is the usual practice to treat the turbulent dispersion terms [contained within the divergence term of Equation (1)], using the so-called "gradient-transport" theory, where the turbulent flux is expressed as a product of a pseudo diffusivity and a mixing-ratio gradient; that is,

$$\begin{array}{c}\text{turbulent species flux}\\ \text{in i direction}\end{array} = -c_{air}K_{iAm}\frac{\partial r_{Am}}{\partial x_i},$$ (7)

where r_{Am} (note the distinction between r_{Am} and r^*_{Am}) is the mixing ratio of m-bound A (moles of m-bound A per mole of air). This allows the expression of the x, y, and z velocity components as

$$U_{Am} = U_{air} - \frac{K_{xAm}c_{air}}{c_{Am}}\frac{\partial r_{Am}}{\partial x},$$ (8)

$$V_{Am} = V_{air} - \frac{K_{yAm}c_{air}}{c_{Am}}\frac{\partial r_{Am}}{\partial y},$$ (9)

and

$$W_{Am} = W_{air} - \frac{K_{zAm}c_{air}}{c_{Am}} \frac{\partial r_{Am}}{\partial z} \mid W_{Am}. \tag{10}$$

Here the air-velocity components can be interpreted in terms of the air-velocity vector

$$\underline{v}_{air} = U_{air}\underline{i} + V_{air}\underline{j} + W_{air}\underline{k}. \tag{11}$$

w_{Am} (note the distinction between w_{Am} and w_{Am}^*) is the nondiffusive vertical movement (e.g., precipitation fallspeed) of m-bound species A, relative to the vertical air motion.

Two important points concerning Equation (7) should be noted. The first of these is that this practice of expressing turbulent transport in terms of a diffusion coefficient and a spatial gradient is a highly approximate one, and represents a strong "assumption" in the mathematical development. More elegant closure approximations have been suggested [77], but their applications have been discouraged in practice because of their additional complexity. The second noteworthy point involves the use of the mixing ratio, r_{Am}, rather than the concentration, c_{Am}, to describe the driving force. Such practice results in a more thermodynamically appropriate expression of the driving force, although it has not been followed in many presently existing air-quality models. This can become of great practical importance in the vertical dimension, where density variations are pronounced and inappropriate results can arise from concentration-based expressions.

Substituting Equations (8) through (11) into Equation (1) provides the counterpart equation set

$$\frac{\partial r_{Am}}{\partial t} = -\underline{v}_{air} \cdot \nabla r_{Am} - \frac{1}{c_{air}}\left[\frac{\partial}{\partial z}(c_{air}r_{Am}W_{Am}) \right.$$

$$\left. + \nabla \cdot (c_{air}\underline{K}_{Am} \cdot \nabla r_{Am}) + w_{Am}^* + r_{Am}^* \right], \tag{12}$$

$$A = 1, 2, \ldots, \text{total number of species},$$

$$m = 1, 2, \ldots, \text{total number of media},$$

where $\underline{K}_{Am}$ is the diffusivity tensor representing the diagonal components K_{xAm}, K_{yAm}, and K_{zAm}.

Equation (12) is a specialized form of Equation (1), and incorporates the assumptions of
● gradient-transport theory [cf. Equation (7)], and
● negligible changes in air density (c_{air}) in the x- and y-directions.
This less general form is often applied directly for model formulation, although Equation (1) has experienced extensive direct application as well. Several simple examples of these equations' use are presented in a later section.

Counterpart *integral* material-balance forms can be formulated simply by defining a spatial domain and formally integrating appropriate transport and generation terms over this volume. For example, behavior associated with a storm

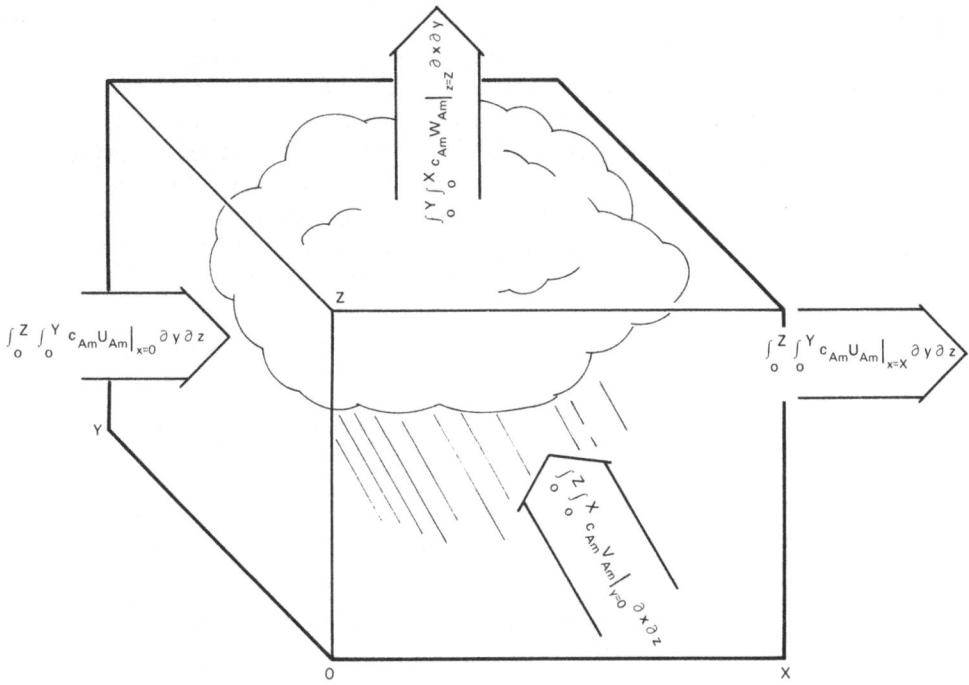

Fig. 12. Schematic of an integral material balance

existing within a box X, Y, Z (cf. Fig. 12) can be expressed as

$$\frac{\partial}{\partial t} \int_0^X \int_0^Y \int_0^Z c_{Am} \, \partial z \, \partial y \, \partial x$$

$$= \int_0^Y \int_0^Z (U_{Am} c_{Am}|_{x=0} - U_{Am} c_{Am}|_{x=x}) \, \partial z \, \partial y$$

$$+ \int_0^X \int_0^Z (V_{Am} c_{Am}|_{y=0} - V_{Am} c_{Am}|_{y=Y}) \, \partial z \, \partial x$$

$$+ \int_0^X \int_0^Y (W_{Am} c_{Am}|_{z=0} - W_{Am} c_{Am}|_{z=z}) \, \partial y \, \partial x$$

$$+ \int_0^X \int_0^Y \int_0^Z w^*_{Am} \, \partial z \, \partial y \, \partial x$$

$$+ \int_0^X \int_0^Y \int_0^Z r^*_{Am} \, \partial z \, \partial y \, \partial x. \qquad (13)$$

Equations (1), (12), and (13) can be considered the material-balance corner-stones for the vast majority of precipitation-chemistry calculations. Special applications of these forms will be described later in this chapter.

Energy Balances

Normally temperature-field information is required prior to solving Equations (1), (12), or (13). Such information is necessary to define temperature-sensitive variables such as pollutant concentrations and reaction rates. In addition, those models involving cloud and precipitation formation require detailed temperature input to govern the delicate thermodynamic balances associated with water condensation and ice formation. If temperature fields are not available as model input data, then they must be generated via solution of an *energy balance*.

A *differential* energy-balance counterpart to Equation (1) is:

$\frac{\partial}{\partial t}(c_{air}C_vT) =$		Time rate of change of sensible energy in a unit volume of atmosphere =
$-\nabla \cdot (c_{air}C_v\underline{v}_{air}T)$		$-$ rate of loss of sensible energy by flow through walls of volume element
$-p(\nabla \cdot \underline{v}_{air})$		$-$ rate of work done by system by expansion (14)
$+\Gamma$		$+$ rate of sensible heat gained via latent heating effects
$+\tilde{R}$		$+$ rate of sensible heat gained by radiant heating effects,

where T represents absolute temperature in degrees Kelvin.

This is a somewhat specialized form of the general energy equation, and is based upon the assumptions that air is an ideal, inviscid gas. Here p is the local air pressure, C_v is the molar specific heat of air at constant volume, and Γ and $\tilde{R}$ account for sensible heat gain from latent heat and radiant-heat effects, respectively. Radiant heating will be ignored throughout the remainder of this discussion since its effect can be considered negligible for most practical purposes. The latent-heating term Γ can cover a variety of mechanisms. For the present we will presume that this term can be segregated into

- vapor-liquid phase changes,
- vapor-solid phase changes, and
- solid-liquid phase changes;

thus

$$\Gamma = \lambda_c r_c^* + (\lambda_c + \lambda_f)r_d^* + \lambda_f r_f^*, \tag{15}$$

where

- λ denotes latent heat of phase transformation (molar basis),
- r* denotes rates of phase transformation (molar basis), and
- c, d, f, pertain to condensation, vapor deposition to ice, and freezing, respectively.

One should note that under conditions of a one-dimensional steady state involving clear air, this equation reduces to the well-known expression for a dry-adiabatic temperature profile:

$$\gamma_d = \frac{dT}{dz} = -\frac{M_{air}g}{C_p g_c} \tag{16}$$

g being the local acceleration of gravity, g_c the universal gravitational conversion factor, M_{air} the molecular weight of air, and C_p the molar specific heat of air at constant pressure.

If one were to apply gradient-transport theory to the energy equation in a manner similar to that employed above for the material-balance Equation (12), i.e.,

turbulent sensible heat flux in i-direction
$$= -c_{air}C_pK_{x_iH}\frac{\partial T}{\partial x_i}; \quad x_i = x \text{ or } y \text{ direction}$$

$$= -c_{air}C_pK_{zH}\left(\frac{\partial T}{\partial z} - \gamma_d\right); \quad z \text{ direction}$$

then

$$\frac{DT}{Dt} = W_{air}\gamma_d + \frac{1}{c_{air}}\nabla \cdot [c_{air}\underset{\sim}{K}_H \cdot (\nabla T - \gamma_d)]$$

$$+ \frac{1}{c_{air}C_p}[\lambda_c r_c^* + (\lambda_c + \lambda_f)r_d^* + \lambda_f r_f^*]. \tag{18}$$

As with Equation (12), the application of gradient transport theory to this equation results in a significant degree of approximation.

It is important to note that the energy-balance equation is typically coupled to material-balance equations for water, since the latent-heat term is linked to the phase-transformations of clouds, rain, and snow. For this reason many of the more advanced scavenging calculations deal with simultaneous heat and mass transfer. One particularly important form of the energy balance is based on the presumption that, within this simultaneous process, the balance between water vapor and cloud water is always at equilibrium; thus the conservation equation for cloud-water mass can be combined with Equation (18) to give:

$$\left(1 + \frac{r_{wv}\lambda_c^2}{C_pRT^2}\right)\frac{DT}{Dt} = \frac{\lambda_c^2}{c_{air}RC_p}\nabla \cdot \left(\frac{c_{air}r_{wv}}{T^2}\underset{\sim}{K}_{wv} \cdot \nabla T\right) + \frac{r_{wv}\lambda_c\gamma_dW_{air}}{RT}$$

$$- \frac{\lambda_c}{c_{air}}\frac{\partial}{\partial z}\left(\frac{c_{air}r_{wv}\gamma_dK_{zwv}}{RT}\right) + W_{air}\gamma_d$$

$$+ \frac{\lambda_f}{c_{air}C_p}(r_d^* + r_f^*)$$

$$+ \frac{1}{c_{air}}\nabla \cdot [c_{air}\underset{\sim}{K}_H \cdot (\nabla T - \gamma_d)]. \tag{19}$$

Equations (14), (18), and (19) can be considered as energy-balance corner-stones, which are direct counterparts to the differential material-balance forms (1) and (12). *Integral* energy equations could be formulated in an analogous fashion; such forms have not experienced common useage in scavenging calculations, however, and thus will not be discussed further in this chapter.

Momentum Balances

Material-balance equations (1), (12), and (13) also require wind-field input prior to their solution. In a manner analogous to that for temperature fields, such data may be supplied from an external source. If such a source does not exist, however, then the information must be generated internally. Under such circumstances one must solve some form of the momentum equation

$\dfrac{\partial \varrho \underline{v}_{air}}{\partial t} =$		Time rate of change of momentum in a small volume element, per unit volume =
$-\nabla \cdot \varrho \underline{v}_{air} \underline{v}_{air}$		— rate of loss of momentum by transport through walls of volume element, per unit volume,
$-\nabla p$		— pressure forces acting on element, per unit volume, (20)
$-\underline{F}_v$		— viscous forces acting on element, per unit volume,
$+ \varrho \underline{g}$		+ gravitational force acting on element, per unit volume.

Very few current scavenging models employ internal solutions of the momentum equation, and we shall not elaborate upon it at this point. It should be noted, however, that the material-, energy-, and momentum-balance equations represent a mutually coupled set, and that the complexity of any particular scavenging model usually increases drastically with the number of equations included.

A final and extremely important point is the recognition that the above "cornerstone" equations represent a rather general reference set, which can be groomed in a formal systematic manner to create specific models of precipitation-chemistry behavior. One could, alternatively, formulate simplified property balances directly from simplified conceptual models, and avoid dealing with the complex forms shown above. This should be strongly discouraged. Systematic elimination of terms from the general equations enforces a realization of model assumptions, whereas the alternate approach often tends to obscure assumptions to the point where inadvertant errors arise. Several examples of model formulation, via simplification of the above equations, will be given in a later section.

Precipitation-Scavenging Modeling:
Description of Physicochemical Microprocesses

Introduction

As was noted in the context of Fig. 11, the scavenging model's governing equations must be supplied information regarding the "microscopic" phenomena involving chemical reaction and mass exchange between physical phases of the system. These microscopic features are, for the most part, consolidated in the terms w^*_{Am} and r^*_{Am} of Equations (1) and their counterparts. In the present subsection we shall deal with these interphase transport and reaction terms progressively, starting with the below-cloud capture of nonreactive aerosols. Several descriptive formulae concerning these processes will be presented, for convenient access in future modeling endeavors.

Elementary Raindrop Statistics

Before embarking on detailed mechanistic discussions it is appropriate to mention briefly the importance of size-distributed raindrop and particle systems. Typically the elements of such systems range over a variety of sizes, and this size-dispersion can have rather profound influences on the scavenging process.

Usually raindrop spectra are characterized mathematically by a probability-density function $f_R(R)$, defined such that the total volume occupied by rain, in a given unit volume of atmosphere, at any time, is [1]

$$V_{rain} = \frac{4\Pi N_T}{3} \int_0^\infty R^3 f_R(R)\,dR,$$
(21)

where N_T is the total number of raindrops residing in the unit volume and R is the radius of any given drop.

Raindrops fall at different velocities, which depend on their sizes. If these velocities are described by the function $v_z(R)$, [2] then extension of the above definition gives the following equation for rainwater volume falling through a unit horizontal area per unit time:

$$J_{rain} = -\frac{4\Pi N_T}{3} \int_0^\infty R^3 v_z(R) f_R(R)\,dR.$$
(22)

J_{rain} (units of l/t or $l^3/l^2 t$) is the normally measured "rainfall rate."

Dispersion of raindrop sizes implies a variety of ways for describing pollutant concentrations associated with the rain phase. If the concentration of species A in size-R raindrops is $\hat{c}_{Arain}(R)$ (quantity of A per unit volume of water), then the concentration of A in a unit volume of the total space (air + water) containing the ensemble of drops is

$$c_{Arain} = \frac{4\Pi N_T}{3} \int_0^\infty R^3 \hat{c}_{Arain}(R) f_R(R)\,dR.$$
(23)

If all the rain existing in the volume were captured and mixed, the resulting "mixed-average" concentration (quantity of A per unit volume of water) would be

$$\hat{C}_{Arain} = \frac{4\Pi N_T}{3 V_{rain}} \int_0^\infty R^3 \hat{c}_{Arain}(R) f_R(R)\,dR$$

$$= c_{Arain}/V_{rain}.$$
(24)

---

1 The assumption of spherical raindrops is employed here and throughout the remaining chapter.
2 $v_z(R)$ is defined as positive upward here, to conform with a Cartesian coordinate system.

If one were to collect all the raindrops falling through a unit horizontal area and measure their mixed concentration, $\hat{C}_{Arain}$, then

$$\hat{C}_{Arain} = -\frac{4\Pi N_T}{3J_{rain}} \int_0^\infty R^3 v_z(R) \hat{c}_{Arain}(R) f_R(R) dR. \tag{25}$$

The reader should keep these distinct definitions of concentration in mind as he reads the remainder of this section. In particular, he should note that c_{Arain} (quantity of rainborne A per unit volume of *total space*) is the entity appropriate for use as a dependent variable in basic governing Equation (1). Extensive use will be made of Equations (21) through (25) in the following text.

Below-Cloud Scavenging of Nonreactive Aerosols: Homogeneous Aerosols

We shall begin by considering the simple case of a homogeneous aerosol, that is, one which is composed of particles of uniform size. This can be analyzed most conveniently by visualizing a volume element of air as shown in Fig. 13. If a single raindrop falls through this element, one can define a *collection efficiency* in terms of the total aerosol existing in the pathway of the drop, and the amount actually collected during the raindrop's passage through the element; that is,

$$E(R, a) = \frac{\text{mass of particles collected during}}{\Pi R^2 \Delta zm}, \tag{26}$$

where R is the raindrop's projected radius, a is the (effective) aerosol particle radius, and m is the mass of particles per unit volume existing within the element prior to the drop's passage.

From Equation (26) and Fig. 13, it is obvious that the accumulation rate of particle mass by the falling drop should be

$$\Pi R^2 v_z m E(R, a),$$

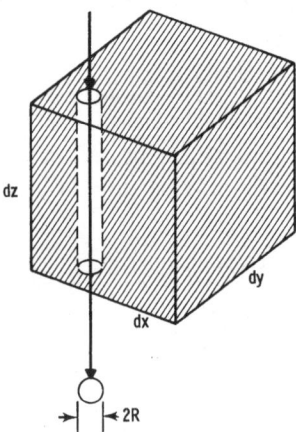

Fig. 13. Schematic of raindrop passing through a polluted volume

where v_z is the vertical velocity of the raindrop (negative downward). Now for an ensemble of raindrops falling through the element, whose size distribution is characterized by the probability-density function $f_R(R)$, the total rate of pickup can be derived by integration over the total range of raindrop sizes. This total pickup rate is by definition equal to the interphase transport rate of aerosol *to* the air *from* the drop [w^*_{Am} in Equation (1)], thus:

$$w^*_{Aair} = \Pi N_T m \int_0^\infty R^2 v_z(R)\, E(R,a)\, f_R(R)\, dR .$$ (27)

This relationship also can be expressed in terms of a *washout coefficient*, Λ, defined as

$$w^*_{Aair} = -\Lambda c_{Aair} .$$ (28)

From Equation (27),

$$\Lambda = -\Pi N_T \int_0^\infty R^2 v_z(R)\, E(R,a)\, f_R(R)\, dR ,$$ (29)

relating the washout coefficient to the efficiency.

Often one does not possess specific knowledge concerning the raindrop size-distribution, and wishes to parameterize Λ in terms of either V_{rain} or J_{rain}. This can be accomplished in an approximate sense if one assumes some general form for $f_R(R)$ and formally integrates (29) and (21) or (22) to provide the functional relationship

$$\Lambda = F_1(V_{rain}) = F_2(J_{rain}) .$$ (30)

Many of the models summarized later in the chapter utilize this parameterized approach.

From Equations (28) and (29) one can in principle compute the desired interphase-transport rate, w^*_{Aair}, if the entities E, v_z, and f_R are known; these will be discussed separately in the following paragraphs.

Aerosol Capture Efficiency, $E(R,a)$: The efficiency term $E(R,a)$ depends upon a host of possible collection mechanisms. These include:
- impaction of aerosol particles on the raindrop,
- interception of particles by the raindrop,
- Brownian motion of particles to the raindrop,
- nucleation of a water drop by the particle,
- electrical attraction,
- thermal attraction (thermophoresis), and
- diffusiophoresis,

and have been discussed at length by numerous previous authors (e.g., Dingle and Lee [20], Hidy [57]). The last three of these mechanisms are of secondary importance in the case of below-cloud scavenging, except for rather special circumstances (cf. Fig. 14). The nucleation mechanism, while potentially significant in many applications, is disregraded in the present context on the presumption that the aerosol in question is hydrophobic, and thus will maintain its fixed particle size, a. Slinn [100] has analyzed the first three of these mechanisms, and has suggested the following three formulae for computing the corresponding component efficiencies:

$$e_{\text{impaction}} = [(S - S_*)/S + C)]^{3/2},\tag{31}$$

$$e_{\text{interception}} = 4a/R,\tag{32}$$

$$e_{\text{diffusion}} = 4Sh/(Re\,Sc).\tag{33}$$

Here the Sherwood Number can be calculated from the Froessling Equation

$$Sh = \frac{2k_y R}{Dc_{\text{air}}} = 2 + 0.6\,Re^{1/2}\,Sc^{1/3},\tag{34}$$

where

$$S\,(\text{Stokes Number}) = \frac{-2a^2\varrho_p v_z}{9R\varrho_a v}$$

$$S_*\,(\text{Critical Stokes Number}) = (1.2 + L/12)/(L + 1)$$

$$Sc\,(\text{Schmidt Number}) = \frac{v}{D}$$

$$Re\,(\text{Reynolds Number}) = -2Rv_z/v$$

$$C = 2/3 - S_*$$

$$L = \ln(1 + Re/2)$$

$$D = \text{molecular (Brownian) diffusivity}$$

$$k_y = \text{mass-transfer coefficient}$$

$$v = \text{kinematic viscosity of air}$$

$$\varrho_a = \text{density of air}$$

$$\varrho_p = \text{density of the aerosol particle}.$$

More refined and involved estimates of these component efficiencies are available in the more recent literature [102].

The corresponding numerical values of E(R, a) obtained by summing Equations (31), (32), and (33) exhibit the well-known tendency to become large for both very large and very small particle sizes, and to become small at intermediate sizes in the range of 0.1 microns (cf. Fig. 14). As indicated by the figure, this minimum region is sometimes known as the "Greenfield Gap" [39], and can be partially "filled" by phoretic mechanisms under specific atmospheric circumstances [106]. Since contributions of these secondary mechanisms are neglected in the present approach, E(R, a) values computed in this manner can be considered to be conservatively *low* estimates of actual behavior. One can, of course, establish a corresponding *upper-limit* estimate of E(R, a) by simply setting it to unity. Since this practice can lead in some cases to efficiency-values three orders of magnitude higher than those obtained from Equations (31) through (33), it is somewhat limited in value – at least in the present context where nucleation is assumed unimportant. Because of this, Equations (31) through (33) are recommended for practical use under these conditions.

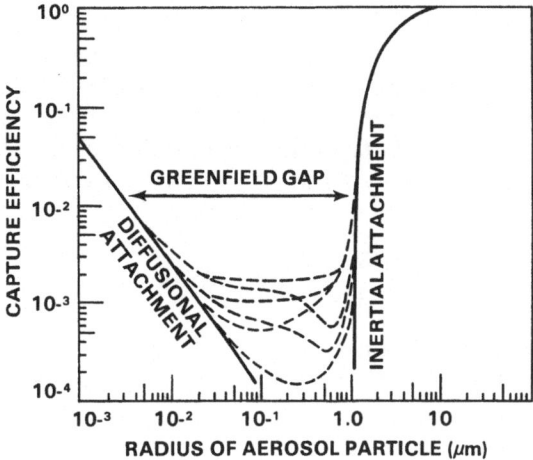

Fig. 14. Schematic of capture efficiency for a single drop size, dotted lines pertain to contributions by various phoretic and electrical mechanisms. Adapted from Wang and Pruppacher [106]

Probability-Density Function for Raindrop Size, $f_R(R)$: Owing to the complexity of rain-formation processes, no really satisfactory formulation exists to describe raindrop size spectra in a totally comprehensive manner. Undoubtedly the most-often applied probability-density function for raindrop size distributions is that of Marshall and Palmer (cf. Pruppacher and Klett [84]):

$$f_R(R) = C_2 \exp(-C_2 R). \tag{35}$$

Here $C_2 = 8.2 \, J^{-0.21} \, mm^{-1}$ is a rainfall-rate dependent parameter (J = rainfall rate in mm/h). It is suggested also in this context that the total number-concentration of raindrops, N_T, should lie close to $980 \, J^{0.21}$ drops/m^3.

Equation (35) is recommended for initial calculations in conjunction with Equations (27) and (29). If more comprehensive computations are desired, one may choose to utilize other types of spectral equations, or employ field measurements of the actual rain spectra for the specific case at hand.

Raindrop Fall Velocity, v_z: Estimation of the fall velocity of raindrops is complicated by the presence of temperature and pressure gradients, and internal circulations and deformations within the drop (cf. Pruppacher and Klett [84]). For practical application, however, empirical fits to measured data provide the most practical means for characterization. The equations of Dingle and Lee [20], given by the forms

$$v_z = 27.2692 - 1206.2884R + 348.0768R^2 \quad (0.05 \le R \ge 0.7 \, mm) \tag{36}$$

and

$$v_z = -155.6745 - 613.4914R + 123.3392R^2 \quad (0.7 \le R \ge 2.9 \, mm) \tag{37}$$

provide a balance between simplicity and accuracy, and are recommended as a starting point for use in below-cloud calculations.

v_z in the above expressions pertains to individual raindrops whose sizes are specified. As noted previously, however, some types of models involve parameterized or "lumped" rain features where the total rainwater volume V_{rain} is known without respect to its drop-size distribution. Such models usually require a representative velocity, characterizing the mass-average of the total rain-drop ensemble; for example, if the local rainwater volume "concentration" is given by Equation (21) then the associated average rain velocity is

$$\hat{V} = \frac{4\Pi V_{rain}}{3} \int\limits_0^\infty R^3 v_z f_R(R)\, dR .\tag{38}$$

Integrated expressions of Equation (38) have been used frequently in the literature. Easter and Hales [24], for example, apply Marshall-Palmer statistics in conjunction with approximations to v_z to obtain appropriate $\hat{V}$ expressions for use in a generalized reactive storm model.

When (as is the usual case) *both* the raindrop and aerosol spectra are size-distributed, an extension of Equation (27) is required. If the aerosol *mass* concentration is described by the probability density function $f_{mass}(a)$, then

$$w^*_{Aair} = \Pi N_T m \int\limits_0^\infty \int\limits_0^\infty R^2 v_z(R)\, E(R, a) f_R(R) f_{mass}(a)\, dR\, da ;\tag{39}$$

$$\Lambda_{mass} = -\Pi N_T \int\limits_0^\infty \int\limits_0^\infty R^2 v_z(R)\, E(R, a) f_R(R) f_{mass}(a)\, dR\, da .\tag{40}$$

Parameterized forms describing Λ_{mass} in terms of V_{rain} or J_{rain} can be derived in a fashion analogous to that of Equation (30).

It should be noted that in Equation (40) Λ_{mass} is defined as a *mass* scavenging coefficient. If one were interested in actual *numbers* of particles washed out, one could define a *number* scavenging coefficient simply by inserting a number-density function in place of $f_{mass}(a)$. Several examples of scavenging coefficient curves for various rain and aerosol spectra are given by Dana and Hales [16]. In all subsequent discussion we shall drop the subscript on Λ_{mass}, upon indicating that Λ is indeed the mass scavenging coefficient unless indicated otherwise.

Below-Cloud Scavenging of Aerosols: Enhancement by Water Condensation

The above discussions of below-cloud scavenging, which were based on the presumption that aerosol particles do not change their sizes during the scavenging process, are somewhat unrealistic. Most common aerosol particles do indeed act as nuclei for water condensation at high humidities (Junge [62]), and appreciable changes in their sizes can be expected to occur as a result. This combined with the rather radical changes in E with particle size predicted from Equations (31) through (33) suggests that considerable modifications of below-cloud scavenging rates can occur via the condensation process.

Sizes and growth rates for nucleated droplets depend on the nature of the nucleating particle and the water-vapor content of its surrounding environment. At high humidities, growth can be very rapid for small droplet sizes; as the droplets

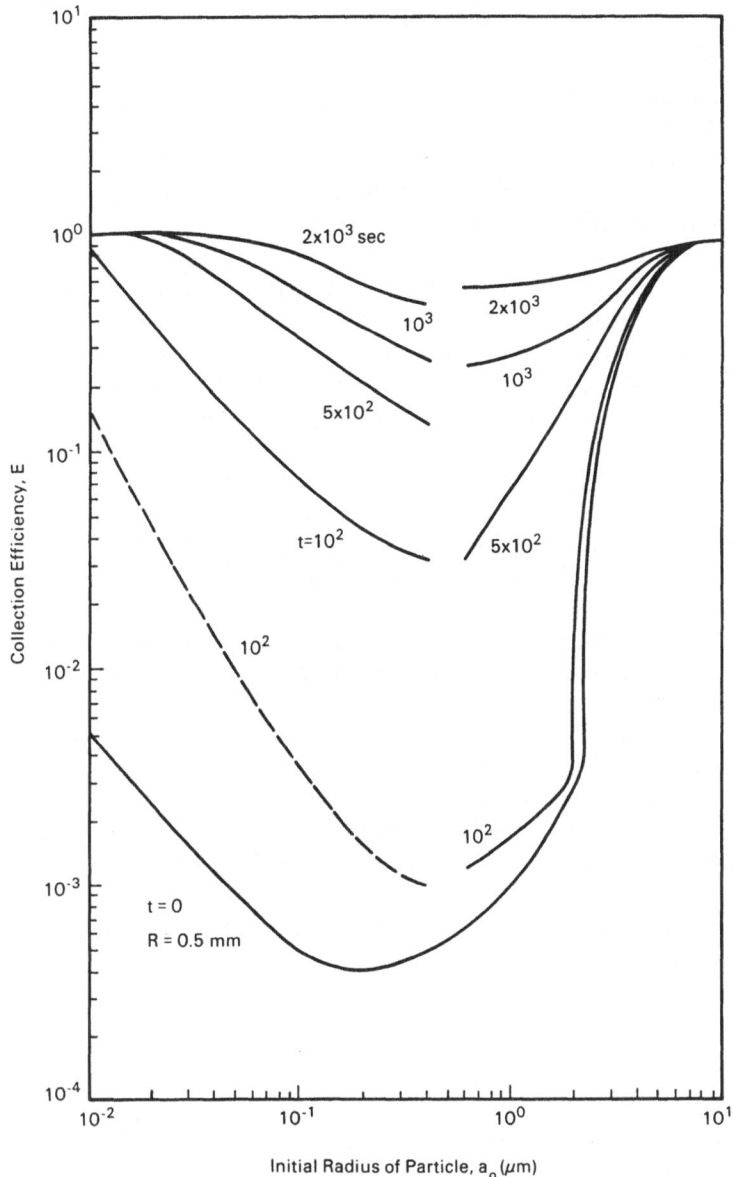

Fig. 15. Theoretical washout efficiencies for 0.5 mm radius raindrops, for nucleating aerosols [100]

become larger, however, the process slows significantly. Given a supersaturation of one percent, for example, a one-micron droplet will double its size via condensation within a few seconds; for a ten-micron particle the corresponding doubling time is of the order of several minutes (cf. Mason [74]).

   Our theoretical capability to deal with the prospect of nucleation and condensational growth in below-cloud scavenging is presently at an unsatisfactory state. Slinn [100] has taken the rather straight-forward approach of:

1. selecting an aerosol particle of dry size a_0,
2. calculating the size of the particle as it grows by condensation, assuming specific growth conditions,
3. calculating revised values of E as a function of time, corresponding to the increasing size of the droplet, and
4. repeating the procedure over a range of dry particle sizes,

to obtain the revised efficiency curves shown in Fig. 15. Here the bottom curve corresponds to a dry aerosol, and is essentially that which would be computed from Equations (31) through (33). The higher curves pertain to washout efficiencies of a growing aerosol, after the indicated growth times.

The efficiency curves on the right-hand side of Fig. 15 are relatively simple, owing to the fact that particles in this size range do not interact significantly with each other via the Brownian diffusion process. They do, however, interact strongly with smaller aerosol particles, and thus the collection efficiencies of the latter are altered appreciably. Slinn has attempted to account for this in preparing the left-hand curves in Fig. 15. His efforts have been limited, however, by the assumptions needed with regard to the characteristics of the large-particle end of the droplet spectrum. This has led to the variety of curves and the discontinuities that appear on the figure.

All of the above uncertainties, plus the generally unknown time-humidity history of an air parcel in a below-cloud scavenging environment add up to the fact that we have very little competence in prediction of below-cloud scavenging rates of aerosols under conditions where nucleation occurs. This effect undoubtedly serves to push scavenging efficiencies in the direction of the upper "asymptote" ($E=1$) condition mentioned previously. Just how effective this process is, however, is understood very poorly. Rather comprehensive analyses of aerosol growth with condensation are available (cf. Mason [74], Fitzgerald [34], Jensen and Charlson [60], Harrison [51], and Johnson [61]), and some fragmentary field measurements of plume scavenging exist (Radke et al. [85]), but much remains to be accomplished before a really satisfactory understanding of this phenomenon is attained.

Below-Cloud Scavenging of Aerosols: Collection by Snow

The irregular and varied geometries of snow particles lead to difficulties in assessment of their size distributions, fall velocities and scavenging efficiencies; thus the computation of below-cloud scavenging by snow emerges as a problem fraught by even more difficulty than that described previously for rain. The usual mathematical approach to this problem is to define some sort of "efficiency," which is comparable to that defined in Equation (26), and is based on an equivalent diameter of one type or another. Slinn [100] suggests:

D_e = diameter of sphere circumscribing the snow particles, and proceeds to express a corresponding washout coefficient by the form

$$\Lambda = -\frac{\Pi N_T}{4} \int_0^\infty D_e^2 v_z(D_e)\, E(D_e, a)\, f(D_e)\, dD \qquad (41a)$$

[cf. Equation (29)]. Combining this with an expression describing equivalent precipitation-rate J in terms of D_e, he proceeds to the simplified form

$$\Lambda = \gamma J E(D_e, a)/D_m,\tag{41b}$$

where γ is a constant of the order of unity, and D_m is a characteristic length scale whose numerical values are summarized in Table 1.

Table 1. Characteristic lengths of ice crystals for use in Equation (41b). From Slinn [100]

Crystal type	D_m (cm)
Graupel	0.014
Rimed plates and stellar dendrites	0.0027
Powder snow and spatial dendrites	0.001
Plane dendrites	0.00038
Needles	0.0019

Slinn also provides a semi-empirical equation for E (not given here), which is similar in form to Equations (31) through (33). An upper asymptote for the system, of course, is simply $E = 1$.

An alternative approach to snow-scavenging calculations, which is based on a more empirical framework, is that outlined by Knutson and Stockham [65]. These authors give explicit expressions for Λ which are functions of J, a, and temperature, and are based upon direct experimental observations.

In comparing the above results as well as the computations and measurements by additional investigators, it becomes readily apparent that several orders of magnitude uncertainty exists in typical applications of snow-scavenging calculations. Much more research needs to be accomplished, especially in the area of physical measurements of E, before a satisfactory computational capability will exist in this area.

Below-Cloud Scavenging of Nonreactive Gases

In the preceding discussion of aerosol scavenging it was assumed tacitly that interphase transport of pollutant between the atmosphere and a falling drop was *irreversible*; that is, once collected the aerosol could not escape back to the air from the aqueous phase. This feature is reflected in Equations (27) and (28), which imply that w^*_{Aair} is always negative, that is, interphase transport should always be *from* the gas phase *to* the drop.

In the case of gases, which can both *absorb in* and *desorb from* water, the irreversibility assumption is generally invalid; and under such conditions it is usually necessary to reformulate expressions for w^*_{Am} which take reversibility into account. This is done most conveniently by discontinuing use of the efficiency concept [as expressed in Equation (26)] and employing instead a corresponding expression for *flux*[3] of pollutant from the falling hydrometeor:

$$F_A = -\frac{K_y}{c_{air}}(c_{Aair} - \hat{c}_{Adrop}/\alpha) \quad \begin{array}{l}\text{moles of A passing from drop's}\\ \text{surface to gas phase, per unit}\\ \text{interfacial area.}\end{array} \tag{42}$$

Here K_y is an overall mass-transfer coefficient, $\hat{c}_{Adrop}$ is the concentration of species A in the drop (moles A per volume of water), and α accounts for the solubility of the gas. One should note that both absorption and desorption are predicted by Equation (42), depending on the relative magnitudes of c_{Aair} and $\hat{c}_{Adrop}/\alpha$. One should observe also that, because of small molecular masses and relatively high diffusivities of gaseous pollutants, diffusion predominates as an interphase transport mechanism.

Diffusive transport in *both* the gaseous and aqueous phases is important in determining gas scavenging rates, and it is usually convenient to consider these effects individually in terms of gas- and liquid-phase coefficients k_y and k_x, such that

$$\frac{1}{K_y} = \frac{1}{k_y} + \frac{1}{k_x\alpha} \tag{43}$$

(cf. Bird et al. [8]). k_y can be estimated from Equation (34). Evaluation of k_x is somewhat more difficult, although for many gases of high or moderate solubility (large α) its relative effect in Equation (43) is small and it can be neglected (cf. Barrie [4]). The reciprocal of the mass-transfer coefficient is commonly interpreted as a mass-transfer "resistance," and this equation is simply an expression of the sum of these resistance components (see Peters [82] for a more detailed discussion).

Hales [44] provides equations for k_x under conditions where no convection occurs internally in the drop, and mixing is by molecular diffusion only. Such expressions depend upon time-exposure history of the raindrop, and tend to be rather complicated, although these can be approximated by the form

$$k_x = 0.278\, D_{Ax}/R, \quad (\text{moles/cm}^2\,\text{s}), \tag{44}$$

where D_{Ax} is the molecular diffusivity of species A in water in cm^2/s, and R is the drop radius in centimeters. Hales argues that these "stagnant-drop" conditions should represent a lower limit for mass transfer and that for all practical purposes natural behavior should be bracketed by calculating Equation (43) with
1) $k_x = \infty$ (gas-phase limited)
2) k_x given by Equation (44)   (stagnant drop).
Garner and Lane [35] suggest that the effect of internal mixing, for large drops at least, can be approximated by simply multiplying Equation (44) by the factor 2.5 and incorporating it with Equation (43). An excellent practical discussion of internal mixing effects is given by Peters [82].

A number of investigators have published elaborate treatments of internal drop circulation (e.g., Walcek and Pruppacher [105]). Usually, however, practical computations do not warrant this level of detail. The simplified scheme outlined above is recommended for all routine scavenging calculations involving gases.

---

3 F_A is defined here as being positive outward from the drop. It is related to w^*_{Arain} by multiplying by drop surface area and integrating over the raindrop spectrum, as shown in Equation (46).

Recently the concept of the *accomodation coefficient*, a_c, has been discussed at some length in the scavenging literature [54, 13], and because of this it should be mentioned briefly at this point. In a somewhat approximate sense the traditional accomodation coefficient can be defined as that fraction of gas molecules colliding with a surface which physically adhere for periods of time that are long compared to the vibrational periods of the surface molecules [18]. Obviously if molecular "bounce-off," corresponding to a_c being less than unity, occurs at a raindrop surface, then the interfacial fluxes predicted by the preceeding approach will be in error. Sherwood, Pigford, and Wilke [97] note that this effect can be expressed in terms of an additional gas-phase mass-transfer coefficient, which may be derived from kinetic gas theory:

$$k_i = 44.4 \, a_c (MT)^{-0.5} \, \text{moles/cm}^2 \, \text{s}, \tag{45}$$

where M is the molecular weight of the absorbing gas and T is the temperature in degrees Kelvin.

This "interfacial resistance" can be incorporated into the overall mass-transfer coefficient by including the reciprocal of k_i in the resistance summation of Equation (43). Since typical k_y values are of the order of 10^{-3}, a_c would have to be less than 10^{-2} to make much difference as a rate-controlling feature.

Accomodation coefficients are difficult to measure, especially for gas-liquid interfaces. Measurements associated with solid interfaces tend to yield values near unity for most gases other than hydrogen and helium, although deviations from this rule definitely do exist [18]. Data associated with liquid surfaces tend to be highly variable and uncertain although the more reliable measurements with the common gases indicate values approaching unity in most cases [97]. At our present state of understanding it is probably acceptable to ignore the influence of k_i for most practical cases in gas-scavenging calculations.

There is an emerging tendency, however, for some authors to confuse the "bounce-off" phenomenon with the normal absorption-desorption reversibility depicted by Equation (42). This usually occurs when some irreversible form of Equation (42) is applied to a reversible situation, and a pseudo accomodation coefficient is applied to compensate for the desorption effect; for example,

$$F_A = -a_c K_y c_{Aair}/c_{air} \quad \text{(incorrect)}.$$

Such practice should be discouraged strongly, because it is a misrepresentation of physical processes and can lead to extremely inaccurate mathematical characterization of the scavenging process.

On the assumption of spherical raindrops, Equation (42) can be integrated to provide a general expression for the interphase transport rate, which is a gas-scavenging counterpart to Equation (27).

$$w_{Aair}^* = \frac{4\Pi N_T}{c_{air}} \int_0^\infty R^2 f_R(R) \, K_y(R) \, (c_{Ay} - \hat{c}_{Arain}(R)/\alpha) \, dR. \tag{46}$$

As noted previously, many modeling applications deal with overall rain features only, and require parameterized expressions for w_{Am}^*. This is usually accomplished

Table 2. Gas-Scavenging coefficients of Levine and Schwartz [70] (based on Marshall-Palmer distribution)

Rainfall rate (mm/h)	Λ_r (s^{-1})
1	0.00012
5	0.00029
15	0.00059
25	0.00076

in terms of a "reversible scavenging coefficient"; i.e.,

$$w^*_{Aair} = -\Lambda_r(c_{Aair} - \hat{C}_{Arain}/\alpha) = -\Lambda_r\left(c_{air} - \frac{c_{Arain}}{V_{rain}\alpha}\right), \tag{47}$$

where $\hat{C}_{Arain}$ is the mixed-average concentration of species A dissolved in the rain existing in the volume element. Λ_r can be defined easily in principle by combining Equations (46) and (47):

$$\Lambda_r = \frac{\dfrac{4\Pi N_T}{c_{air}} \int\limits_0^\infty R^2 f_R(R)\, K_y(R)\, (c_{Aair} - \hat{C}_{Arain}/\alpha)\, dR}{c_{Aair} - \hat{C}_{Arain}/\alpha}. \tag{48}$$

Solution of Equation (48) to achieve numerical values for Λ_r is complicated by the necessity to perform integrations associated with the aqueous-phase concentrations, and normally some sort of simplifying assumption is applied. Levine and Schwartz [70] for example, obtained solutions to Equation (48) for a variety of raindrop spectra, which were based upon the assumption that the factor $\hat{C}_A/\alpha$ is effectively invariant with drop size. Levine and Schwartz's Λ_r values corresponding to a Marshall-Palmer raindrop distribution are shown in Table 2.

As a final point on nonreactive gases, it is important to note that several limiting cases exist which can lead to simplification of the above equations. One such case is that of "equilibrium scavenging," wherein the physical exchange of component A between the gaseous phase and a raindrop occurs so readily that a diffusional equilibrium exists, and

$$\hat{c}_{Arain} = \hat{C}_{Arain} = \hat{\hat{C}}_{Arain} = \alpha c_{Aair}. \tag{49}$$

Under such conditions aqueous-phase concentrations can be calculated immediately from their gaseous-phase counterparts, without resort to complex material-balance considerations.

The existence of a diffusional equilibrium depends upon the diffusional response time of the raindrop relative to the rate of change of its ambient concentration field. A dimensionless criterion has been derived to describe conditions where equilibrium scavenging should exist [44]. This can be expressed as

$$\frac{-3K_y c_{Aair}}{v_z R \left|\dfrac{dc_{Aair}}{dz}\right| \alpha c_{air}} \gg 1, \tag{50}$$

where $\dfrac{dc_{Aair}}{dz}$ is interpreted (conservatively) as the maximum vertical gradient in the concentration of airborne A experienced by the falling drop.

A second limiting case pertains to the situations where component A may be considered so soluble in water that uptake can be considered essentially as an irreversible process. Under such conditions Equation (47) loses its term associated with C_{Arain}, and essentially reverts to the simpler form (28). A dimensionless criterion for the existence of irreversible gas scavenging is

$$\frac{\alpha\hat{c}_{Arain} - c_{Aair}}{c_{Aair}} \gg 1. \tag{51}$$

The above relationships, especially Equation (50), are of large utility in many practical calculations. Additional limiting cases occur in the context of reactive gas scavenging, which is discussed immediately below.

Below-Cloud Scavenging of Reactive Gases

Up to the present point we have focused primarily on the w^*_{Am} term of Equation (1), with little mention of the reaction term r^*_{Am}. Models involving simultaneous solutions of equations containing the dry terms r^*_{Aair}, r^*_{Bair}, ... are commonplace today, and reactive scavenging models are straightforward extentions of these techniques. A first-order decay of rainborne A, for example, can be expressed directly by the form

$$r^*_{Arain} = -kc_{Arain}, \tag{52}$$[4]

where k is the first-order reaction-rate coefficient.

A slightly more complex example, pertaining to the aqueous-phase oxidation of S^{IV} by H_2O_2, is

$$r^*_{SO_2rain} = -\frac{4.6E7\, c_{H_2O_2rain}\, c_{SO_2rain}}{c_{rain}\left[1 + \dfrac{K_1}{\hat{C}_{H^+rain}}\left(1 + \dfrac{K_2}{\hat{C}_{H^+rain}}\right)\right]}, \tag{53}$$

where K_1 and K_2 are dissociation constants for SO_2 [25].

Numerous examples of aqueous-phase reaction expression appear in the published literature (e.g., Gradel and Goldberg [37], Hoffman and Calvert [59], Peters [82], Schwartz [91]), and these features will not be discussed at length here. It should be noted in this context, however, that much of our present uncertainty in scavenging modeling is a direct consequence of our lack of knowledge of aqueous-phase reaction processes. Much remains to be accomplished in this important research area.

There are times when the w^*_{Am} and r^*_{Am} terms behave in ways which tend to simplify modeling calculations. In other cases extended and rigorous computations are required. These situations can be grouped into four alternative classes, which are described immediately below:

---

4 This equation is based upon the assumption that the raindrop ensemble-average concentration is an acceptable "driving force" for characterizing the chemical reaction.

Class 1: Rapid Reversible Reactions

If the chemical reaction is *rapid* and *reversible* with a nonvolatile product, i.e.,

$$A \underset{k_r}{\overset{k_f}{\rightleftarrows}} B \tag{54}$$

then the scavenging interaction usually can be treated as a *pseudophysical absorption process* (cf. Sherwood and Pigford [96]). With this treatment the nonreactive gas-scavenging approach characterized by Equations (42) and (43) can be utilized directly for calculation, as long as an appropriate means for describing solubility is available.

Sulfur dioxide, which dissociates rapidly in water to form hydrogen and bisulfite ions, is a good example of a substance adhering to this behavior. A dimensionless criterion describing conditions acceptable for pseudophysical absorption calculations is

$$\frac{R k_r \alpha c_{air}}{3 K_y} \gg 1, \tag{55}$$

where k_r is the reverse rate coefficient depicted in Equation (54).

Class 2: Rapid Irreversible Reactions

If the chemical reaction is rapid, irreversible, and leads to a non-volatile product, then mass transfer to the raindrop's surface usually can be considered as the rate-limiting step in the scavenging process. Under such conditions the system reverts to an irreversible scavenging situation, and can be treated in exactly the same manner as the case described above for irreversible physical scavenging. A dimensionless criterion for irreversible conditions in reactive scavenging systems is

$$\frac{R k \alpha c_{air}}{3 K_y} \gg 1, \tag{56}$$

where k is the first-order rate coefficient for the irreversible reaction.

Class 3: Decoupled Reactions

With relatively slow chemical reactions (e.g., ozone oxidation of SO_2 to form sulfate), two modes of below-cloud scavenging may be isolated. These correspond to the nonreactive absorption of airborne gas and reactive depletion of dissolved gas within the drop. If relaxation times for the absorption step are short compared to those for reaction, then the first of these modes may be treated as an isolated "nonreactive" scavenging process in a quasi-independent manner. Likewise, the reactive mode can be simplified under some circumstances to allow a relatively straightforward calculation to be performed. Again, dimensionless criteria may be derived to describe conditions where such "decoupling" assumptions are allowable [44].

Class 4: General Case

In the more general case, involving the possibility of multiple reactions, competitive effects, or volatile reaction products, one usually has little choice other than formulating a detailed mathematical description of the mass-transfer –

chemical-reaction process. Generalized numerical frameworks for below-cloud scavenging calculations of this class have been reported. (Drewes and Hales [22], Hales et al. [43]). Computations for specific reactive systems in well-mixed environments have been presented by Overton et al. [80], Hill and Adamowicz [58], and Durham et al. [23].

In-Cloud Scavenging of Gases and Aerosols

Common mechanisms contribute to the scavenging process regardless of whether or not it occurs within a visible cloud system, and thus the distinction between in- and below-cloud scavenging is somewhat artificial. There is, however, a definite shift in the relative importance of these mechanisms. Readdressing the collection pathways itemized previously it seems obvious that, for cloud environments where condensation is occurring, *nucleation* should play a much more dominant role in contacting particulate pollutant with condensed water. Also, because of the importance of evaporation-condensation cycles in typical cloud systems, *electrical, thermal,* and *diffusiophoretic* forces should be expected to become relatively important (cf. Dingle and Lee [20]), although nucleation is often presumed to dominate as a first approximation. Interception and impaction, on the other hand, can be expected to become insignificant for the attachment of primary pollutant particles to cloud droplets, although they definitely remain important as mechanisms of accretion of pollutant-laden droplets to falling hydrometeors.

Aerosol nucleation in cloud environments can be considered as an in-cloud extention of the water-enhanced process of below-cloud scavenging, which was described previously. It is difficult to quantify the interphase transport rates associated with this process, however, owing to the complex nature of cloud fluctuations, supersaturation effects, and competition for water vapor between potential nuclei. Numerous elaborate computations of this process have been published (e.g., Scott [90], Jensen and Charlson [60]); many generalized scavenging models, however, simply express the nucleation component of w_{Am}^* as a pseudo first-order rate process, or as a phenomenon which is constrained by the number of cloud-droplets in the vicinity. This is an area where large uncertainty exists. A much more solid understanding is badly needed, since this question has strong implications on important issues such as the "nuclear winter" and nonlinearities in the acid-precipitation formation process.

Since so many things can happen simultaneously in a cloud environment, it is somewhat difficult to disaggregate various classes of behavior, as was done previously for the case of below-cloud scavenging. Because of this, it is useful to consider the totality of in-cloud processes simultaneously, in the context of "interaction" diagrams as shown in Figs. 16a and 16b. Figure 16a is a simple interaction diagram for water, which shows water being transferred between various classes via a diversity of mechanistic pathways. These pathways can be considered as rate and/or equilibrium processes as appropriate, and although quantification of these terms is beyond the scope of this text, it is not difficult to envision how one would derive mathematical equations characterizing each one of these mechanisms. It isn't difficult, either, to envision how one could formulate a

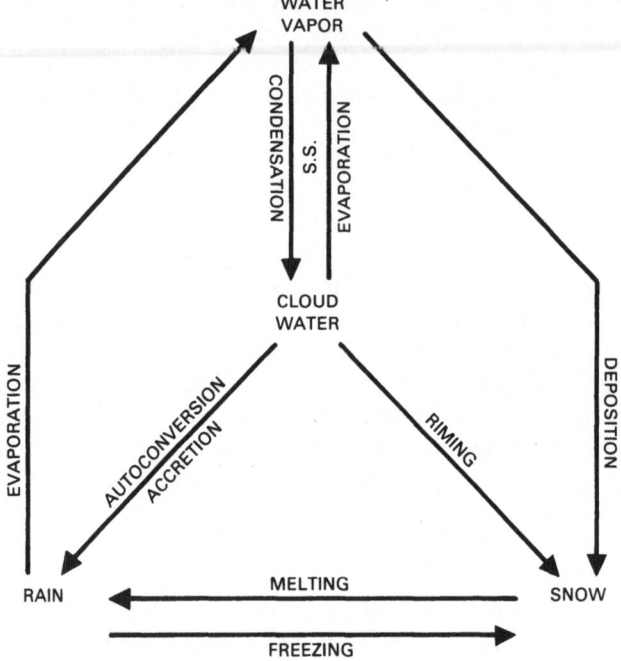

Fig. 16a. Interaction diagram for water classes

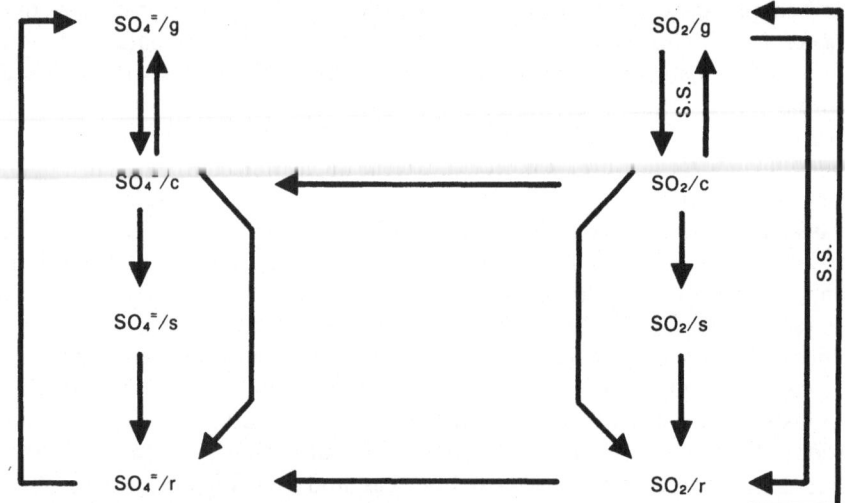

Fig. 16b. Interaction diagram for sulfur classes

precipitating-cloud model by incorporating the various w^*_{Am}'s expressed by these equations into differential material balances of the form (1).

Figure 16b provides an interaction diagram for sulfur compounds, which complements the water interaction diagram above. This figure can be used as a basis for formulating rate expressions for sulfur transformations, in a manner similar to that described above for water. In addition to dealing with *physical* rate terms (w^*_{Am}), however, this diagram also indicates *chemical* rate processes (r^*_{Am}) associated with the oxidation of SIV to SVI.

It is important to note the cross-linkages between the mechanisms of Figs. 16a and 16b. Physical conversion of cloud-borne sulfate ion to rainborne sulfate ion, for example, depends on the same accretion and autoconversion processes that convert cloud water to rain. Thus this conceptual scheme, and its mathematical counterpart, provide a convenient and straightforward way of modeling the intimate linkages that occur between cloud physics and scavenging.

At present there are only a few scavenging models which are based on the above approach [24, 25, 40, 55], and it has been common practice to consolidate all the w^*_{Am} and r^*_{Am} terms into some sort of lumped parameter such as a storm-average scavenging coefficient, or a scavenging ratio (e.g., Slinn [100], Scott [90]). These consolidated-coefficient approaches, however, share the disadvantage that they possess little mechanistic definition or versatility. The more general interaction-diagram – based methods, although more burdensome computationally, are much more amenable to interpretive efforts. They also provide straightforward means for incorporating laboratory measurements of wet-chemical transformation rates directly into the scavenging models, and thus are of great utility in linking laboratory, theoretical, and field efforts under a practical framework.

Because of the highly interactive nature of reactive storm systems, it is appropriate to terminate our focused discussion of micro-physiocochemical processes at this point in favor of a more global treatment. This will be pursued in the following section, which addresses combination of the conservation equations with their micro-physiocochemical components to provide complete models of the precipitation-chemistry formation process.

Combination of Governing Equations and Microprocess Descriptions: Example Model Integrations

Introduction

The previous two sections of this chapter have addressed individually the topics of property-balance – derived governing equations, and their physicochemical microcomponents. The present section deals with the merger of these two entities into complete modeling equations, and examines techniques for their solution. A wide variety of model equations exists; some of these may be solved analytically while others require numerical approximation and machine computation. We shall begin here with simple examples, and progress to later discussions of more computationally intensive systems.

Differential Material Balances and Below-Cloud Scavenging

Irreversible Scavenging of a Nonreactive Pollutant
Through a Uniformly Polluted Air Mass

This is an extremely simple situation. Air temperature is basically ignored or, at best, presumed to be invariant over the computational domain, so that no solution of the energy equation is required. Furthermore, the assumed uniformity of pollution concentrations effectively eliminates the concentration-derivative terms in Equation (1); this, in turn, eliminates dependence on wind velocity, so a solution of the momentum equation is not necessary.[5] Two expressions of Equation (1) are required, however, one pertaining to pollution in rain and the other to pollution in air.

Following the formal "grooming" process described earlier, we first address the gaseous-phase version of Equation (1). As noted above, the velocity-related terms in this equation are zero. Furthermore, there is no reaction term, thus the equation reduces to

$$\frac{dc_{Aair}}{dt} = w^*_{Aair} = -\Lambda c_{Aair} \tag{57}$$

[cf. Equation (28)], which becomes

$$c_{Aair} = c_{Aair}|_{initial} \exp(-\Lambda t) \tag{58}$$

upon integration, giving the temporal behavior of the airborne pollution as the scavenging process progresses in time.

It should be noted that we have not applied an aqueous-phase counterpart of Equation (1) to the problem at this point. Certainly this could be accomplished if one were interested in spatial profiles of rainborne A. Normally, however, one wants only information concerning precipitation chemistry at the ground, which may be computed directly from a simple integral material balance. This will be treated later, when integral balances are discussed specifically.

It is imperative that the reader appreciate that Equations (57) or (58) are NOT definitions of the scavenging coefficient, Λ. They are, rather, expressions of the differential material balance and its integrated product under highly specialized conditions. Substantial errors have been made in the past through neglect of this distinction.

Reversible Scavenging of a Nonreactive Pollutant
Through a Uniformly Polluted Air Mass

It was possible to employ a single equation in the previous example primarily because the model system was *decoupled*, in the sense that Equation (57) contains no terms associated with the aqueous-phase concentration. For reversible scavenging, however, w^*_{Aair} depends upon both the gaseous- and aqueous-phase

---

5 In this example and the next several examples we shall avoid both the energy and momentum equations by assuming that T and v_{air} are either previously known or else unimportant to the problem-at-hand.

concentrations; i.e.,

$$w^*_{Aair} = -\Lambda_r\left(c_{Aair} - \frac{c_{Arain}}{V_{rain}\alpha}\right). \tag{47}$$

This, when combined with the mass-conservation equation for airborne A under the stipulated conditions, gives

$$\frac{dc_{Aair}}{dt} = -\Lambda_r\left(c_{Aair} - \frac{c_{Arain}}{V_{rain}\alpha}\right), \tag{59}$$

which requires information concerning c_{Arain} prior to solution.

In this case aqueous-phase concentrations must be calculated using an aqueous counterpart to Equation (59). From Equation (1) the appropriate form for this is

$$\frac{\partial c_{Arain}}{\partial t} = -\frac{\partial}{\partial z}(c_{Arain}W_{Arain}) + w^*_{Arain}$$

$$\text{(Note that } w^*_{Aair} \text{ in this case } = -w^*_{Arain}), \tag{60}$$

or

$$\frac{\partial c_{Arain}}{\partial t} = -\frac{\partial}{\partial z}(c_{Arain}W_{Arain}) + \Lambda_r\left(c_{Aair} - \frac{c_{Arain}}{V_{rain}\alpha}\right)$$

here the z-component of the divergence term must be included with the equation because, although the air concentrations are distributed evenly in the vertical, the concentrations of rainborne A are not.

Simultaneous Equations (59) and (60) are somewhat difficult to solve analytically.[6] One can, however, impose additional constraints to decouple the system and simplify the problem. If one presumes the scavenging process to occur so slowly that c_{Aair} does not change appreciably, for example, then

$$c_{Aair} = \text{constant in time} \tag{61}$$

and

$$\frac{\partial c_{Arain}}{\partial t} = 0 = -W_{Arain}\frac{\partial c_{Arain}}{\partial z} + \Lambda_r\left(c_{Aair} - \frac{c_{Arain}}{V_{rain}\alpha}\right) \tag{62}$$

which may be integrated to obtain

$$c_{Arain}(z) = \alpha c_{Aair}V_{rain} - (\alpha c_{Aair}V_{rain} - c_0)\exp\left(\frac{\Lambda_r(z-z_0)}{W_{Arain}V_{rain}\alpha}\right), \tag{63}$$

where c_0 is a reference concentration of A in rain at some height z_0.

Irreversible Scavenging Through a Nonreactive Gaussian Plume

As with the initial example, the irreversible nature of this problem allows gas-phase equations to be formulated and solved independently from their aqueous-phase counterparts. The standard "Gaussian plume equation" is simply a solution to the gas-phase form of Equation (1), based upon the assumptions of steady-state

---

6 Solution of these simultaneous equations is certainly possible, and can be accomplished in a straightforward manner using Laplace transforms. The reader is encouraged to obtain this solution as an exercise.

behavior, an elevated point source, zero w^*_{Am} and r^*_{Am}, and constant transport properties. If this governing equation is modified to include w^*_{Aair} as expressed by the irreversible Equation (28), the solution is [98].

$$c_{Aair}(x, y, z) = \frac{Q}{2\Pi\sigma_y\sigma_z U} \exp\left(\frac{-y^2}{2\sigma_y^2}\right)\left[\exp-\left(\frac{(z-h)^2}{2\sigma_z^2}\right) + \exp-\left(\frac{(z+h)^2}{2\sigma_z^2}\right)\right]$$

$$\exp\left(-\frac{\Lambda x}{U}\right),\tag{64}$$

where Q and h are plume's source strength and release height, U is the wind velocity, and σ_y and σ_z are the plume spread parameters.

Aqueous-phase concentrations may be computed by formulating and solving an aqueous version of Equation (1). If one again assumes constant, vertical rainfall, this equation becomes

$$-W_{Arain}\frac{dc_{Arain}}{dz} + \Lambda c_{Aair} = 0.\tag{65}$$

The average rainborne pollutant concentration c_{Arain} can be calculated at any point x, y, z simply by inserting Equation (64) into Equation (65) and integrating. In particular, c_{Arain} at ground level[7] is

$$c_{Arain}(x, y, 0) = \frac{\Lambda Q}{\sigma_y W_{rain} U \sqrt{2\Pi}} \exp\left(\frac{-y^2}{2\sigma_y^2}\right)\exp\left(\frac{-\Lambda x}{U}\right).\tag{66}$$

Reversible Scavenging Through a Gaussian Plume

In this example we shall make the same "decoupling" assumption that was applied in the previous example of irreversible scavenging; that is, we shall assume that scavenging has negligible impact on the gaseous-phase plume, at least for moderate increments of time. Under such conditions the gaseous plume equation reverts to the normal Gaussian form, $\left[\text{Equation (64) without the } \exp\left(-\frac{\Lambda x}{U}\right)\text{ term}\right]$.

Corresponding aqueous-phase equations can be obtained using the noted formal procedure of simplifying Equation (1) in an appropriate manner, and inserting the reversible Equation (47) to describe w^*_{Am}. For assumed constant, vertical rainfall the result (for ground-level rain concentration) is

$$c_{Arain}(x, y, 0) = \frac{-Q\xi V_{rain}}{2\sqrt{(2\Pi)}\sigma_y U}\exp\left(-\frac{y^2}{2\sigma_y^2} + \frac{\sigma_z^2\zeta^2}{2}\right)$$

$$\cdot\left\{\exp(\zeta h)\,\text{erfc}\left(\frac{-\sigma_z^2\zeta - h}{\sigma_z\sqrt{2}}\right)\right.$$

$$\left. + \exp(-\zeta h)\,\text{erfc}\left(\frac{-\sigma_z^2\zeta + h}{\sigma_z\sqrt{2}}\right),\right.\tag{67}$$

---

7 The reader is again cautioned to observe the differences between c_{Arain}, $\hat{C}_{Arain}$, and the collected-average rain concentration.

where

$$\zeta = \frac{\Lambda_r}{\alpha v_z V_{rain}} \tag{68}$$

and

$$\xi = \frac{\Lambda_r c_{air}}{v_z V_{rain}} \tag{69}$$

(cf. Hales et al. [48]).

In addition to various decoupling assumptions, the above examples have been largely contingent on the presumption of a lumped raindrop spectrum. In the cases involving reversible scavenging the implicit assumption of constant α also has been applied. As noted previously, size-distribution of raindrops can be an important factor, and several authors have extended the above considerations to deal with this situation. These efforts usually have taken the tack of subdividing the raindrop spectra into a number of size-classes, formulating and solving mass-conservation equations for each size, and combining the solutions to predict collected-average behavior at the ground. Usually these modeling efforts have applied decoupling assumptions similar to those described above, and most have resorted to numerical approximation and machine computation, rather than analytical solutions. Examples of such modeling efforts for Gaussian Plumes include those by Hales et al. [48], Barrie [4], Drewes and Hales [22], Hales et al. [43], and Walcek and Pruppacher [105]. Examples pertaining to uniform air masses are given by Adamowicz [1] and Overton et al. [80]. An elegant solution which is analytical and does *not* decouple the gaseous and aqueous-phase equations is given by Slinn [101]. By allowing the equation coupling to remain intact Slinn is able to estimate plume repositioning and distortion arising from the scavenging process.

Scavenging of Reactive Pollutants

Mathematical description of chemically reactive systems may be accomplished by applying the same formal procedure as that described for nonreactive pollutants. Here, however, the term r^*_{Am} is nonzero (at least for some of the species-medium combinations), and must be included with Equation (1) in the grooming and solution process. As an example, the rain-phase species-conservation equation, corresponding to a spatially uniform gaseous-phase pollutant and an aqueous-phase pollutant decay rate given by

$$r^*_{Arain} = -kc_{Arain} \tag{70}$$

would be

$$\frac{\partial c_{Arain}}{\partial t} = -\frac{\partial}{\partial z}(W_{Arain} c_{Arain}) + w^*_{Arain} - kc_{Arain}. \tag{71}$$

Several computer models of reactive below-cloud scavenging have been published. Many of these models (e.g., Overton et al. [80], Hill and Adamowitz [1] have dealt with specific forms for r^*_{Am}, while others (e.g., Drewes and Hales [22], Hales et al. [43]) have been constructed to allow modular replacement of various r^*_{Arain} expressions. Typically these models subdivide the raindrop-size spectrum into multiple classes and make various types of decoupling assumptions. The reader is referred directly to these references for further information on this subject.

Differential Balances and In-Cloud Scavenging

The previous case studies have been based strongly on the assumption of steady precipitation with uniform spatial density. The precipitation field was taken as a "given," and no particular attention was paid to the manner in which it was created.

Such assumptions become more and more difficult to justify as the vertical domain of the modeled system extends upwards into the clouds. Under such conditions the precipitation fields can NOT be considered invariant in the vertical, since the cloud region is precisely where the precipitation is formed. Moreover, as noted in the context of Fig. 16, the precipitation-formation processes are strongly linked to the scavenging phenomena, themselves; this suggests that, in order to simulate scavenging accurately, one must first simulate the cloud processes. Because cloud processes depend on flow fields and energy exchange, this in turn suggests that momentum- and energy-balance equations must be included in the composite description.

The material presented in this subsection summarizes a variety of techniques which have been applied to deal with this composite problem. This presentation begins with techniques which are based soley on pollutant material balances, and then expands into areas where energy and momentum balances have been applied. As with the below-cloud simulations discussed previously, the order of complexity increases as the discussion proceeds.

There are two basic methods for dealing with the composite problem of in-cloud scavenging. The first of these is to create a model which attempts to simulate the interacting processes specifically; the second method is based upon the presumption that some sort of lumped parameters can be defined, which can be applied superficially and yet contain the essence of the composite system. We shall deal with this second, "lumped-composite" method first.

One of these lumped approaches, utilized by many of the earlier models, simply consolidates w^*_{Am} into a single term, which is expressed by an empirical (or at best, semiempirical) scavenging coefficient; that is

$$w^*_{Aair} = -\Lambda c_{Aair},\tag{28}$$

which, upon application with Equation (1), gives

$$\frac{\partial c_{Aair}}{\partial t} = -\nabla \cdot (c_{Aair} \underline{v}_{Aair}) - \Lambda c_{Aair} + r^*_{Aair}.\tag{72}$$

In applying Equation (72) one presumes that the wind fields are known; thus a momentum equation is unnecessary. All of the chemistry, physical attachment, and cloud-process linkages are assumed consolidated into the term Λ, and therefore neither an energy balance nor an explicit description of microprocesses is required. To derive quantitative results from the model, one simply solves Equation (72) subject to appropriate values of Λ. If one desires concentrations of precipitation-borne material, he can proceed to formulate and solve an aqueous-phase counterpart of Equation (72), or a corresponding integral material balance. Several techniques for estimating Λ were noted previously, and numerous applications of this modeling technique appear in the literature (e.g., Baum et al. [7], Watson et al. [108], Lange and Knox [67]).

Most Λ-based in-cloud scavenging models have been formulated with the assumption that Λ does not vary with height, and thus represents a spatially averaged property.[8] This poses a problem in the sense that some regions of cloud systems are known to be much more efficient pollutant extractors than others, and this "differential attrition" is not reflected by the model. This, plus the noted inability to characterize mechanistic features explicitly, seriously detracts from the capability of Λ-based models and has resulted in their increasing unpopularity as more modern techniques have evolved.

A second type of "lumped-composite" model deserves some mention at this point, although it fits more appropriately under the integral material-balance category. This is the so-called "scavenging-ratio" concept, based on the entity

$$\zeta = \text{scavenging ratio} = \frac{\hat{C}_{A\text{precipitation}}}{c_{A\text{air}}}. \tag{73}$$

Usually, although not always, the scavenging ratio is defined in terms of concentrations observed at ground level.

As with Λ-based models, scavenging-ratio models avoid explicit use of the momentum and energy equations by presuming that wind fields are known,[9] and that all precipitation variability and micro-physicochemical linkages are suitably consolidated into the single parameter, ξ. The usual procedure is to formulate a clear-air version of Equation (1) which does NOT contain an interphase transport term, that is,

$$\frac{\partial c_{A\text{air}}}{\partial t} = -\nabla \cdot (c_{A\text{air}} \underline{v}_{A\text{air}}) + r^*_{A\text{air}}. \tag{74}$$

This equation can be solved for a single time increment, and the corresponding ground-level precipitation-borne concentration can be obtained from Equation (73). Using an integral material balance the computed gas-phase concentration can be back-corrected, or "depleted" to account for losses by the scavenging process.

The scavenging-ratio concept has some theoretical appeal, and some rather involved formulations of this parameter have been presented (e.g., Scott [90]). Moreover, scavenging ratios are relatively easy to incorporate into simple models, and consequently they have experienced a great deal of application. They do, however, suffer from the same drawbacks as Λ-based models, regarding spatial and mechanistic specificity. A general formulation of the scavenging-ratio concept will be presented in a later section.

Almost all present-generation regional pollution models have incorporated various forms of the lumped-composite parameterizations described above (cf. Table 3). The primary reason for this, of course, is their outward simplicity. As indicated in the preceding discussion, ζ and Λ can be incorporated with the basic governing equations in a highly straightforward and convenient manner. Deter-

---

8 Note the discussion of spatially-averaged Λ's in the context of integral material balances in the next subsection.

9 This is something of an overstatement. As with a number of other simple applications the wind fields are presumed to exhibit rather idealized behavior in space and in time, which leads to a mathematically tractible situation.

mining reliable values for these variables, however, is difficult at best. This plus the noted problems associated with spatially varying and mechanistically specific behavior has prompted some of the newer regional models to pursue more detailed approaches; these will be discussed in the following text.

Models specifically simulating interacting physicochemical microprocesses have begun to emerge within the past ten years, and promise to increase in importance as our understanding of atmospheric phenomena expands. In essence, these models simulate processes such as those depicted in Fig. 16 directly. For material-balance – only models, such simulations present something of a problem because many of the pathways depend directly on cloud-physics processes, which in turn are strongly contingent on flow fields and energy considerations. If one can assume, however, that wind, cloud, and precipitation fields are known, then it is possible to formulate expressions for rate processes such as those depicted in Fig. 16.

This has considerable practical applicability. At least two developing regional pollution models, for example, implement techniques for estimating vertical cloud and precipitation profiles. These derived profiles are then applied with interaction-diagram – based "scavenging modules" to compute wet removal and precipitation chemistry in a direct and straightforward manner. Other, more diagnostically oriented functions include the application of material-balance – only simulations, using the output of dynamic storm models as a source of wind- and storm-field input. All of these efforts are in an early development stage at the time of this writing, but can be expected to experience expanded application during future years.

Provided that wind fields are known, an energy balance [cf. Equation (14)] can be incorporated to provide the cloud and precipitation fields necessary for mechanistically specific simulations. This can be accomplished in practice without too much difficulty, and several examples of this type of model have been discussed in the literature [24, 55, 40]. As with their material-balance – only counterparts, these types of models can be expected to experience increased use during future years.

Models including momentum balances with material and energy balances, and thus predicting wind fields in addition to the other processes, have seen limited practical use, although examples do exist [49]. This limitation stems primarily from the difficulty of formulating, bounding, and solving momentum equations, and the associated demands on computational resources. In this context it is usually far more practical to formulate and solve a dynamic storm model which is decoupled from the scavenging model, and then use the derived output as a "driver" for the scavenging simulation in the sense described above. This will undoubtedly be the preferred direction of most modeling endeavors during the next ten years.

Integral Material Balances

Integral material balances have been applied extensively for a variety of practical scavenging calculations. It has been the usual practice, however, to derive the associated mathematical equations somewhat informally, performing simple balances around "boxes" or other model volumes on a more-or-less ad hoc basis. Formalized derivation procedures, involving the systematic grooming of Equation (13) in the same manner that Equation (1) was adapted for differential systems, possess the advantages of improved human perception and increased reliability; accordingly, these procedures are strongly recommended for all practical applications.

This subsection illustrates the use of integral material-balance methods by applying Equation (13) to formulate mathematical expressions describing a number of wet-removal systems. As with the differential balances, the intent here is to give the reader a basic idea of these procedures and set a basis for further application. The case studies begin with a rather simple example and progress to more involved scientific applications.

Irreversible Scavenging Through a Uniformly Polluted Air Mass

This is the same physical situation that was described previously in the context of differential material balances, and is an almost trivial application of integral balances owing to the spatial invariability of the gaseous-phase concentration field. To formulate an integral balance one simply places an imaginary box, of dimensions $\Delta X, \Delta Y, \Delta Z$, around the model domain. Focussing first on the gaseous-phase pollutant, one notes that the spatial invariance of c_{Aair} in the present example allows replacement of the integrals with simple products of ΔX, ΔY, and ΔZ. Flow velocities are either zero or spatially invariant, and chemical reaction is neglected; thus Equation (13) becomes

$$\frac{\partial c_{Aair}}{\partial t} \Delta X \Delta Y \Delta Z = w^*_{Aair} \Delta X \Delta Y \Delta Z = - \Lambda c_{Aair} \Delta X \Delta Y \Delta Z \qquad (75)$$

or

$$\frac{dc_{Aair}}{dt} = - \Lambda c_{Aair}$$

which is identical to the form (57) derived from the differential balance, Equation (1).

In the case of rainborne pollutant concentrations, c_{Arain} varies with height, and Equation (13) becomes

$$\frac{\partial}{\partial t} \int_0^z c_{Arain} \, \partial z \Delta X \Delta Y = W_{Arain} c_{Arain}|_{z=0} \Delta X \Delta Y - W_{Arain} c_{Arain}|_{z=z} \Delta X \Delta Y$$

$$+ \int_0^z w^*_{Arain} \, \partial z \Delta X \Delta Y . \qquad (76)$$

If one presumes that the time derivative is small compared to other terms (a quasisteady state) and rainborne concentration at the top of the model domain is zero, then

$$W_{Arain} c_{Arain}|_{z=0} = - \int_0^z w^*_{Arain} \, dz = - \int_0^z \Lambda c_{Aair} \, dz = - \Lambda c_{Aair} \Delta Z \qquad (77)$$

or

$$c_{Arain}|_{z=0} = -\frac{\Lambda c_{Aair}\Delta Z}{W_{Arain}}.$$ (78)

giving an expression for rainborne concentrations of pollutant A at the Earth's surface.[10]

Irreversible Scavenging Through Non-Uniform Air Masses

The strict requirements of the above example can be relaxed somewhat via the definition of spatially averaged parameters. One could, for example, set forth a spatially averaged equation defining the domain of Fig. 12 simply by dividing Equation (13) by the domain's volume; i.e.,

$$\frac{\partial}{\partial t}c_{Am} = \frac{1}{\Delta X}(\overline{U_{Am}c_{Am}}|_{x=0} - \overline{U_{Am}c_{Am}}|_{x=X})$$

$$+ \frac{1}{\Delta Y}(\overline{V_{Am}c_{Am}}|_{y=0} - \overline{V_{Am}c_{Am}}|_{y=Y})$$

$$+ \frac{1}{\Delta Z}(\overline{W_{Am}c_{Am}}|_{z=0} - \overline{W_{Am}c_{Am}}|_{z=Z})$$

$$+ \overline{w_{Am}^*} + \overline{r_{Am}^*}.$$ (79)

If one were to define a spatially averaged scavenging coefficient in the following manner

$$\overline{w_{Aair}^*} = -\bar{\Lambda}\,\overline{c_{Aair}},$$ (80)

then Equation (79) would allow one to approximate gross chemical features of the storm provided, of course, that the initial concentrations, the fluxes, and $\bar{\Lambda}$ are known.

A few examples of modeling efforts leading to storm-averaged scavenging coefficients for nonreactive aerosols are available. Slinn [100], for example, begins with a rather general form of Equation (1), inserts mathematical terms to describe the appropriate physical microprocesses, and averages over space and particle size to obtain the form

$$\bar{\Lambda} = \frac{JE}{2R_m},$$ (81)

where R_m is the volume-mean raindrop size at ground level. The efficiency E is a rather complicated expression reflecting particle attachment and droplet-growth behavior, and is predicted to vary with time. Although very few data exist to test Slinn's expression, tracer-release experiments have been shown to fit Equation (81) reasonably well with an E value of 1/3.

A second expression for a storm-averaged aerosol scavenging coefficient has been derived by Klett [64] in his analysis of the scavenging of nuclear debris. Basically this author has assumed complete attachment of pollutant to cloud

---

10 Note the distinctions between rainborne concentrations, which are given in Equations (23)–(25).

droplets of known size-distribution and has integrated expressions for droplet capture by accretion to obtain the following form:

$$\bar{A} = 4.2 \times 10^{-4} \, E \, J^{0.79} (s^{-1}). \tag{82}$$

Here E is an efficiency term and J represents rainfall rate in millimeters per hour. Klett suggests a numerical value for E of 0.83, based upon his computations describing the cloud-droplet accretion process. Although described here primarily on the basis of integral balances, Equations (81) and (82) may be applied with differential models as well, *provided* that one is willing to accept the conditions associated with their formulation. Present-generation regional models (particularly Lagrangian formulations) tend to minimize the scientific detail of their scavenging calculations and often use such forms to advantage.

Scavenging Ratios

Equation (13) can be groomed easily to give a description of basic scavenging-ratio theory. For this situation one situates a model domain so as to encompass a precipitating cloud as shown in Fig. 12. If one also assumes

- a quasisteady state (no time derivative),
- two-dimensionality (the box is aligned such that V_{Aair}, V_{air}, and V_{Arain} are negligible),
- no spatial variability of the gaseous-phase concentration field, and
- no chemical reaction,

then the following reduced form of Equation (13), describing the behavior of rainborne pollutant, is obtained:

$$W_{Arain} C_{Arain}|_{z=0} \Delta X \Delta Y = - \int_0^z w^*_{Arain} \partial z \Delta X \Delta Y. \tag{83}$$

A similar equation, corresponding to a material balance for rain, itself, can be formulated as well. This is

$$W_{rain} C_{rain}|_{z=0} \Delta X \Delta Y = - \int_0^z w^*_{rain} \partial z \Delta X \Delta Y. \tag{84}$$

Now it is expedient to express the bulk interphase-transport terms on the right-hand sides of the above equations in terms of overall "extraction efficiencies," ε,[11] for pollutant and water vapor entering the model volume; that is,

$$\int_0^z w^*_{Arain} \partial z \Delta X \Delta Y = \varepsilon_{Arain} U_{Aair} C_{Aair}|_{x=0} \Delta Y \Delta Z; \tag{85}$$

$$\int_0^z w^*_{rain} \partial z \Delta X \Delta Y = \varepsilon_{rain} U_{watervapor} C_{watervapor}|_{x=0} \Delta Y \Delta Z. \tag{86}$$

Applying the following substitutions and approximations

$$U_{Aair} = U_{watervapor} = U_{air}$$

$$W_{Arain} = W_{rain}$$

$$C_{Arain} = \hat{C}_{Arain} V_{rain} = \hat{\hat{C}}_{Arain} V_{rain}$$

$$C_{rain} = \hat{C}_{rain} V_{rain}$$

---

11 For these purposes the extraction efficiency is defined as the amount of material delivered to the surface in rain, divided by the total amount of airborne material advected into the storm.

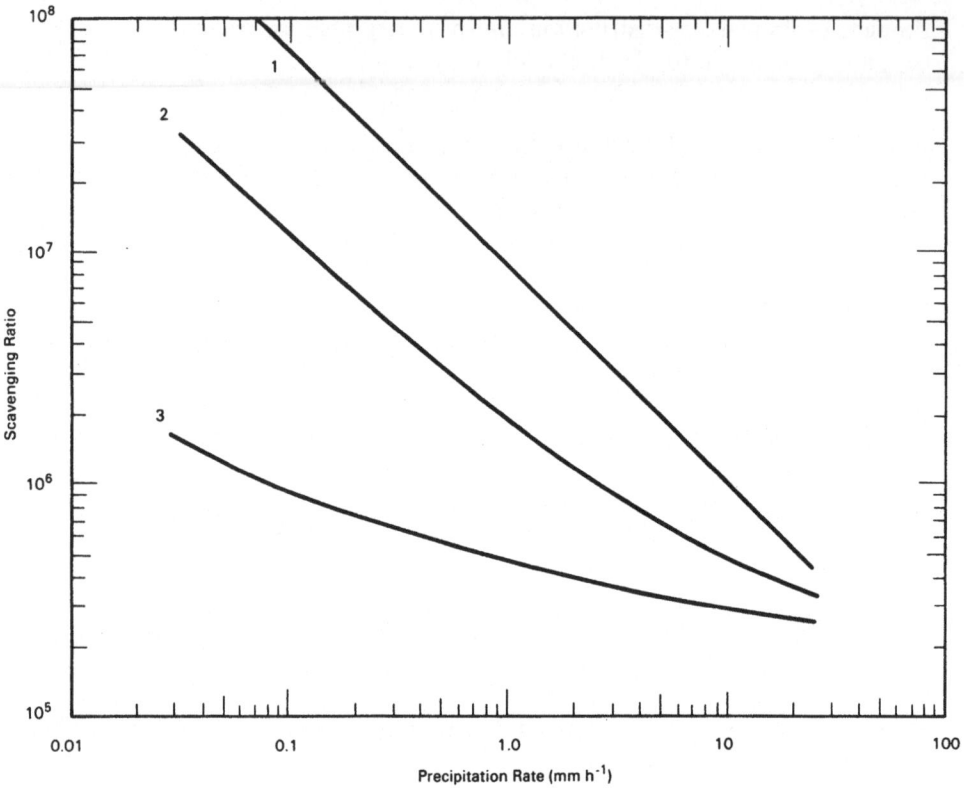

Fig. 17. Estimated aerosol scavenging ratios of Scott [90]. 1) Convective storms; 2) warm frontal storms; 3) storms where Bergeson ice-formation process dominates

results in the important form

$$\xi = \frac{\hat{\hat{C}}_{Arain}}{C_{air}} = \frac{\hat{C}_{rain}\varepsilon_{Arain}}{C_{watervapor}\varepsilon_{rain}},$$ (87)

where $\hat{C}_{rain}$ is the molar density of rainwater (1/18 moles/cubic centimeter).

Several features of this discussion should be noted. First it should be recalled that this integral-balance formulation was applied previously in the context of differential models. Such mixing of model types is not uncommon in practice; however, it often is helpful for the scientist to be fully aware of this fact when such situations do, indeed occur. A second point is that the formulation of Equation (87) has been used to examine several aspects of the physical behavior of (e.g., Englemann [30]). These arguments have been based largely on the presumption that, owing to the intimate interactions between scavenging and precipitation formation (as characterized, for example by the schematics in Figs. 16a and 16b), a strong relationship must exist between the efficiencies ε_{Arain} and ε_{rain}. Scott [90] has applied some rather detailed cloud-physics calculations to estimate this efficiency ratio for aerosol scavenging, and has derived the curves appearing in Fig. 17. These can be utilized as practical (although approximate) indicators of aerosol scavenging for a number of storm types.

Finally, the variety of assumptions leading to Equation (87) should be noted. Several attempts have been made to improve this formulation by applying spatially varying concentration fields, providing for chemical reactions, including reversible behavior, and so forth. While such formulations do show considerable promise as practical indicators, they tend to lack the direct diagnostic power of the more mechanistically specific differential models. Scavenging ratios can be expected, however, to experience continued and even expanded use in practical assessment applications.

Integral Material Balances and Statistical Models

Integral material balances provide an important basis for a variety of statistical scavenging models, and this final subsection on integral balances shall examine briefly the fundamental relationships between these two. Usually the formulations of statistical models contain some rather severe assumptions regarding material-balance behavior. Slinn [99] for example, illustrates this connection by starting essentially with Equation (79) and making the following stipulations:
1. no pollution sources exist within the model domain,
2. no pollutant transport occurs in the y-direction, or across the top of the model domain,
3. the system exists in a steady state, and
4. pollution is not removed by chemical reaction.

By writing an appropriate form of Equation (79) representing the sum of the pollutant A in all possible media, and taking limits in the x-direction, Slinn obtains the equation

$$\frac{\partial}{\partial x}(Uc_A) = \frac{1}{\Delta z}Wc_A|_{z=0} \tag{88}$$

which reduces to the form

$$\frac{\partial c_A}{\partial x} = -\frac{(v_d + v_w)c_A}{U\Delta Z} \tag{89}$$

upon making the further assumption of constant wind speed, and stipulating that the wet- and dry-deposition rates should be directly proportional to the concentration c_A, as dictated by the coefficients v_d and v_w.

Equation (89) an be integrated to obtain the solution

$$c_A(x) = c_A(0)\exp\left(-\frac{(v_d + v_w)x}{U\Delta Z}\right), \tag{90}$$

which, under the above rather restrictive assumptions, can be converted to a Lagrangian reference frame by the substitution $\Delta x = U\Delta t$:

$$c_A(t) = c_A(0)\exp\left(-\frac{(v_d + v_w)t}{\Delta Z}\right) = c_A(0)\exp\left[-t\bigg/\left(\frac{1}{\tau_d} + \frac{1}{\tau_w}\right)\right]. \tag{91}$$

c_A in Equation (91) may be identified as the concentration of pollutant in a parcel of air which has drifted over the Earth's surface for a time-period t. τ_d and τ_w are the characteristic times for the pollutant's removal by dry and wet deposition, respectively, and the pollutant's characteristic lifetime is given by

$$\frac{1}{\tau} = \frac{1}{\tau_d} + \frac{1}{\tau_w} = \frac{v_d}{\Delta Z} + \frac{v_w}{\Delta Z}. \tag{92}$$

Statistical scavenging models typically deal with pollutant residence-time density functions and their parameters (such as τ), and the above formulations illustrate their fundamental relationship with integral material balances. Many examples of statistical wet-removal models exist in the open literature (e.g., Gibbs and Slinn [36], Grandell and Rodhe [87, 88], Baker et al. [3]).

A detailed description of existing statistical wet-removal models is beyond the scope of this chapter. There are, however, a number of important points regarding these efforts which should be noted here. The first of these is that, in contrast to the implications of the simple formulation given above, the wet-removal process is not well-represented by some sort of constant, time-averaged, e-folding parameter such as τ_w. Precipitation events are essentially stochastic phenomena and, as such, act as statistical forcing functions on the wet-removal process; as a consequence many of the more sophisticated models, while formulated basically from integral balances, apply rather elaborate statistical descriptions of storm durations and intensities, and dry periods between storms. Generally the results of these models suggest that wet-removal lifetime density functions decay much more gradually with time than implied by the simple exponential of Equation (91).

An additional important point is that although the modern statistical models contain substantial levels of sophistication, they are generally contingent on a variety of simplifying assumptions. Most of these relate directly to the integral material balances from which they were formulated, and because of this it is usually helpful to examine systematically the material-balance "roots" of a model as a first step in its evaluation. A common component of most statistical models, for example, is some sort of assumption that the wet removal rate during any given storm event is proportional to the pollutant's concentration; as discussed previously, this places some rather severe restrictions on behavior of the physicochemical microprocesses contributing to the scavenging phenomenon. Because of these restrictions statistical scavenging models have made their strongest contributions in rather general areas, especially on the large spatial and temporal scales where deterministic models have met with limited success. Particularly important products of these models have been the establishment of relationships between temporal and spatial variability, the association of temporal variability with pollutant lifetimes, and global estimates of pollutant residence times in the atmosphere. The reader is referred to the cited references for more detailed discussion of this subject.

Model Selection Techniques

Introduction

The previous sections have addressed the mechanics of formulating precipitation-scavenging models and solving the associated equations to obtain estimates of atmospheric concentrations and delivery rates. Quite often, however, the potential user is not so much interested in *formulating* an original model as he is in *selecting* the best possible one from the profusion of available candidates. The present section focuses on this selection process.

A systematic model-selection process can be divided into four major steps, as follows:
1. evaluation of the practical uses for which the model will be applied;
2. evaluation of the scales, domains, and acceptable averaging times of the problem;
3. evaluation of input-information resources; and
4. final model selection.

These four steps are addressed individually in the following subsections.

Model Uses

Precipitation-scavenging calculations can be applied for a large variety of practical ends, and the optimal model chosen for this purpose depends heavily on its intended use. Because of this it is important for the potential user to make a careful assessment of his specific requirements before embarking on a model-selection process. Some common model applications include:
- predicting the impact on precipitation chemistry of proposed new sources, source modifications, and alternate emission-control strategies;
- predicting long-term precipitation-chemistry trends;
- estimating relative contributions of specific sources to precipitation chemistry at chosen receptor sites;
- estimating transport of scavengable pollutants across political borders;
- estimating air-quality improvements occurring as a consequence of the scavenging process;
- selecting sites for precipitation-chemistry sampling stations;
- designing field studies of precipitation scavenging; and
- elucidating mechanistic behavior of the scavenging process on the basis of field measurements.

After reviewing his intended applications, the user will have gained a strong insight for the remainder of the process. This initial step is especially important when the model selection is performed by a group of individuals or by an organization (particularly a government agency). Communication of this information clearly at the outset usually yields strong efficiency benefits at later points in the selection and application process.

Model Scales, Domains, and Averaging Times

This step of the selection process has been discussed earlier in this chapter, and will be summarized here in terms of an itemized list. In general, the potential user should evaluate his requirements in the following areas:
- spatial resolution requirements;
- temporal resolution requirements;
- spatial domain of the model – i.e., where (geographically) the model is to be applied, and the *physical size* of this region;
- temporal domain of the model – i.e., what lengths of time are to be simulated; and
- conceptual domain of the model – i.e., which areas of Fig. 1 are to be simulated.

As was the case with Step 1, a careful and systematic review of these domain features will aid substantially in clarifying later points in the selection process.

Input Information

The successful outcome of all modeling efforts is contingent on the supply of input data to initialize and bound the system, and to "drive" the simulations. Wind, storm-feature, and temperature data are particularly important requirements as drivers for a number of scavenging models; as discussed previously in the context of property balances, these features must be supplied either from measurements, assumptions, or as output from extended balance calculations. The availability of such input information has a large bearing on whether a scavenging model can be based on material balances only, on energy and material balances, or on the combination of material, energy, and momentum calculations. The decision chart shown in Fig. 18 presents these choices, as a summary to the earlier discussion in this chapter.

Final Model Selection

Upon completion of the three screening steps outlined above, the user is typically faced with a variety of available models from which to choose. Furthermore, many of these models possess considerable flexibility, and often the user must determine how to set these flexible options to mimic physical behavior accurately and thus best suit his purposes at-hand. Although there is no real substitute for experience at this process, one may obtain considerable guidance from a decision-tree

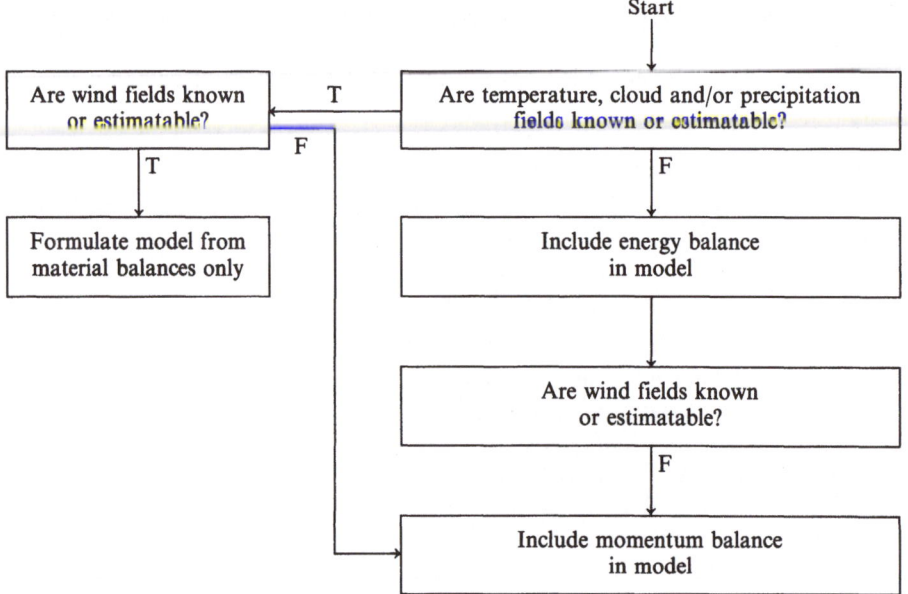

Fig. 18. Preliminary model-scavenging chart

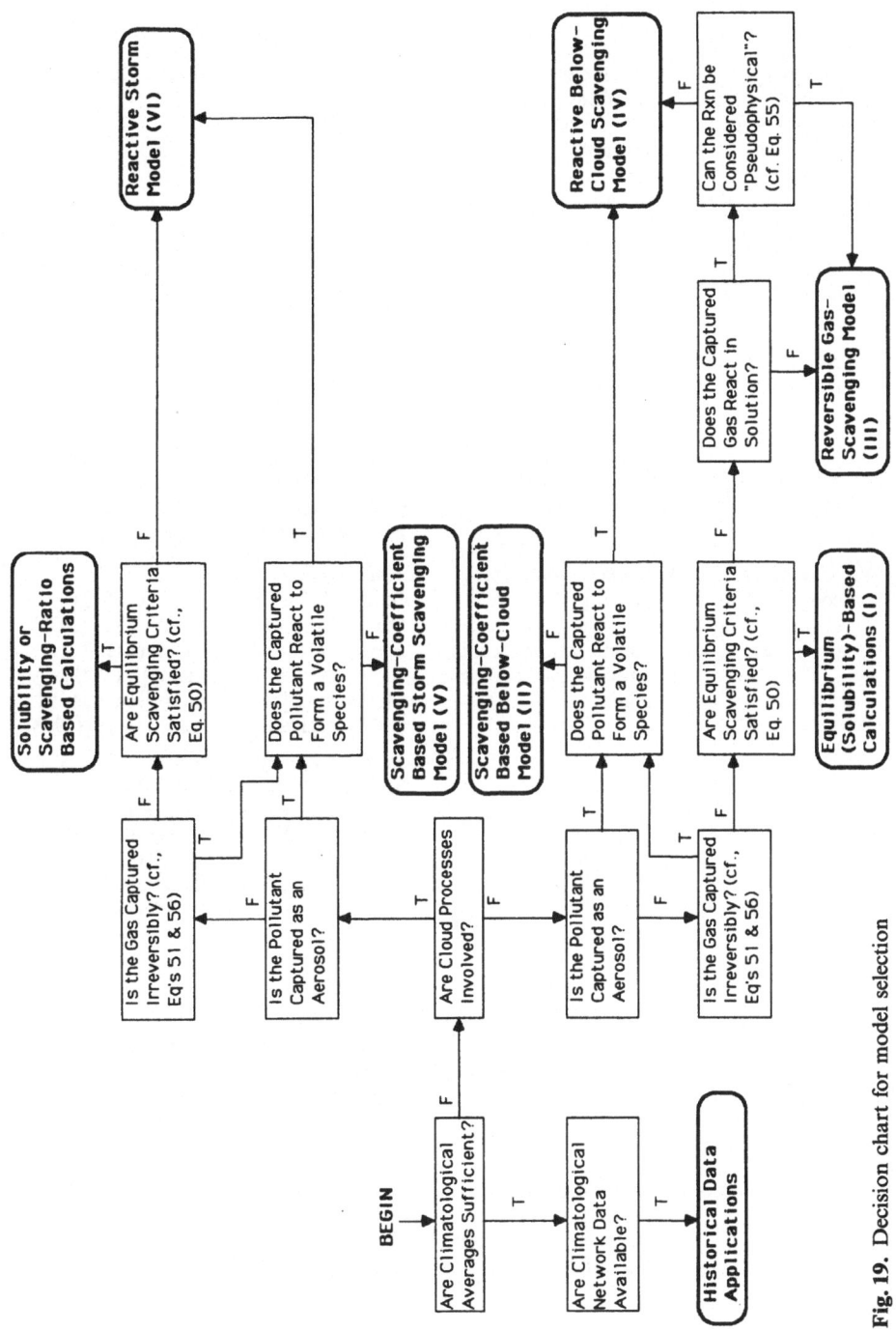

Fig. 19. Decision chart for model selection

Table 3. Selected references on scavenging models

Model category	Application areas	Selected references
I. Solubility-limited ("equilibrium scavenging")	General scavenging calculations for gases of low or moderate solubility in the absence of sharp vertical gradients in the gas-phase concentration	Junge (1963), Davies (1984), Hales and Dana (1982)
II. Scavenging-coefficient (below-cloud)	Highly-soluble gases and nonvolatile aerosols outside of cloud regions	Chamberlain (1953), Engelmann (1968), Fisher (1975), Scriven and Fisher (1975), Wangen and Williams (1978), Dana and Hales (1976), Slinn (1984)
III. Reversible Gas scavenging	Below-cloud scavenging of non-reactive gases (and aerosols, as a limiting case)	Hales et al. (1973), Slinn (1974), Barrie (1978), Walcek and Pruppacher (1984)
IV. Reactive below-cloud scavenging	Below-cloud scavenging of reactive and nonreactive gases and aerosols	Hill and Adamowicz (1977), Adamowicz (1979), Overton et al. (1974), Durham et al. (1981), Drewes and Hales (1982), Hales et al. (1983)
V. Scavenging coefficient or scavenging ratio (in-cloud)	In-cloud or total-storm scavenging or aerosols. Sometimes applied as a parameterization technique for reactive gas-scavenging estimates	*Storm models:* Junge (1963), Dingle and Lee (1973), Storebo and Dingle (1974), Klett (1977), Lange and Knox (1977), Slinn (1977), Scott [I] (1978), Engelmann [I] (1971), Molenkamp [M, E] (1983), Hane [M, E] (1978), Kreitzburg and Leach [M, E] (1978)
		Regional Lagrangian models: Bolin and Persson (1975), Hales et al. (1977), Eliassen (1978), Eliassen and Saltbones (1983), Eliassen et al. (1978), Fisher (1975), Bass (1980), Heffter (1980, 1983), Henmi (1980), Sampson (1980), Bhumralker et al. (1980), Kleinmann et al. (1980), Shannon (1981), McNaughton et al. (1981), Patterson et al. (1981), Voldner et al. (1981)
		Regional Eulerian models: Liu and Durran (1977), Prahm and Christensen (1977), Wilkening and Ragland (1980), Lavery (1980), Lee (1981), Carmichael and Peters (1984), Fay and Rosenzweig (1980)
VI. Reactive storms	In-cloud or total-storm scavenging of reactive and/or nonreactive gases and aerosols	Hegg et al. [E] (1984), Molenkamp [E] (1983), Hales [E] (1982, 1985), Easter and Hales [E] (1982, 1984), Seigneur and Saxena (1984), Heikes and Thompson (1983)

Notes: [I] – integral material balance; [M] – momentum balance included; [E] energy balance included selection technique, such as those which have been presented in the past literature [44].

One such technique is illustrated in Fig. 19. Here the user simply begins at the indicated point on the diagram and attempts to answer the progression of questions, until he finally arrives at the appropriate model class. Answers to some of these questions will be obvious, while guidance on others is given by reference to equations in the preceeding text. Basically this chart can be considered as a summary "road map" to much of this chapter's discussion on model formulation. Selected references corresponding to end-points in Fig. 19 are given in Table 3.

Several general comments pertaining to Fig. 19 are appropriate at this point. The first of these is that the diagram includes both general and specific calculation techniques, and there is a tendency toward increased mechanistic generality (and flexibility) as one progresses from left to right across the chart. Typically the more general models can perform mechanistically specific calculations with at least as much refinement as their more specific counterparts, and thus it is often appropriate to apply models on the right-hand side for calculations assigned to more specific models by the decision tree. Reversible gas-scavenging models, for example, are usually able to handle both scavenging-coefficient and equilibrium simulations, as limiting cases of their normal nonequilibrium, reversible calculations.

A further important comment pertains to multicomponent chemical systems. A comprehensive chemical system normally will include species typified by terminal points throughout Fig. 19, and in such situations one must choose a model of sufficient generality to encompass the entire set. In this context Fig. 19 should be considered not so much as a model-selection tree as a means of selecting the most appropriate methods for setting internal features of the more general models. Finally, Fig. 19's lack of emphasis on scavenging ratios should be noted. Although scavenging ratios represent an important calculational technique their integral nature (with the exception of solubility-limited, "equilibrium" scavenging) tends to give this entity a composite-mechanism cast, which is difficult to characterize in a decision chart. It will suffice at this point to note that scavenging ratios have been applied, on a more-or-less empirical basis, to several of the modeling areas shown on this chart. A number of these models are cited in Table 3.

Model-Measurement Comparisons

Introduction

The preceding sections of this chapter have dealt individually with observed precipitation-chemistry behavior, and the mathematical techniques for its simulation. This final section attempts to provide a partial merger of these two topics by describing the numerical output of selected models, and comparing this directly with observations. In keeping with preceding text, the goal of this section is not to provide a comprehensive analysis of model performance and associated issues. Its intent, rather, is to apply a few illustrative examples to give the reader a general idea of model proficiency as it exists at the present time. More detailed evaluations are the subject of ongoing research, and can be found in more extensive literature references.

Also in keeping with the general thrust of the chapter, the model-measurement comparisons given here will emphasize the characterization of spatial and temporal variability. These comparisons are discussed individually in the following subsections, starting with temporal behavior and progressing to evaluations of spatial distributions.

Within-Event Variability of a Single Storm

A typical time-series of precipitation chemistry within a single event was shown in Fig. 2. During 1982 an intensive study of several such events was completed [15] where, in addition to the precipitation-chemistry data set itself, sufficient information was acquired to initialize and bound a one-dimensional, time-variant, reactive-storm model [25]. This expanded data set included aircraft-measurements of ambient air-pollutant concentrations, balloon-sounding infor-mation, and network weather-monitoring input.

Example model-measurement comparisons are shown in Figs. 20 and 21. The variability of both sulfate- and nitrate-ion concentrations is not dissimilar from that exhibited by Fig. 2, although it should be reemphasized that a variety of time-concentration curves have been observed by studies of this type. It is interesting to note, however, that the time-concentration curves produced as model output show fair agreement with the observations.

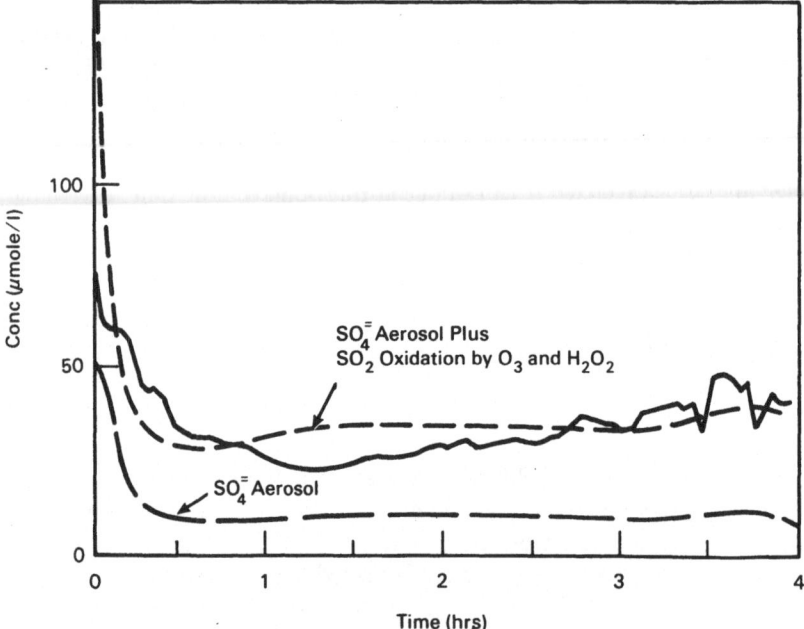

Fig. 20. Comparison of measured within-event rainborne concentrations (solid lines) with predictions from a reactive storm model (dashed line). Data from OSCAR experiment, in northeastern Indiana High-Density Network [15]

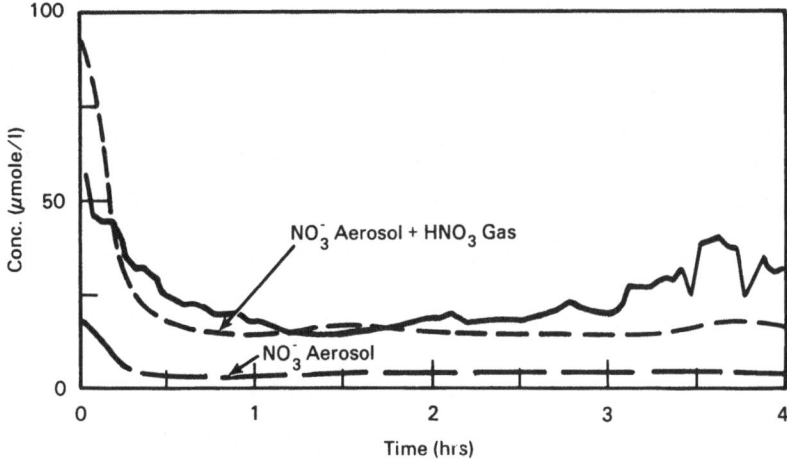

Fig. 21. Comparison of measured within-event rainborne nitrate concentrations (solid line) with predictions from a reactive storm model (dashed line). Data from OSCAR experiment, in northeastern Indiana High-Density Network [15]

The lower of the model-produced curves for sulfate- and nitrate-ion concentrations were computed on the assumption that the scavenging processes involved aerosol capture only. When SO_2 and HNO_3 gas scavenging was incorporated with the calculations the corresponding curves rose appreciably.[12]

It is likely that much of the model-measurement agreement exhibited in Figs. 20 and 21 is fortuitous. Regardless of this, the model's ability to incorporate variability components of the atmospheric source-receptor sequence and translate these to a reasonable variability of precipitation chemistry suggests that basic elements of the modeling process are essentially correct. Major improvements in such simulations can be expected during coming years, as our knowledge of chemical scavenging processes matures.

Long-Term Variability of Regional Precipitation Chemistry

This example contrasts to the above case study in several respects. The model chosen for the present application, known as the ASTRAP code [95], is a *regional* model, and thus simulates all features of Fig. 1 from the sources to the receptors; this is basically different from the reactive-storm code considered above, which was initialized with concentration fields in the storm's vicinity. Moreover, ASTRAP does not incorporate the fundamental scavenging mechanisms used above; rather, this code simply employs a scavenging-ratio concept which has been to some degree force-fit from observations. In this particular example the code has been executed for several years of meteorological data, with an assumed fixed sulfur emission inventory; thus the effect has been to "freeze" the emissions and focus specifically on meteorological components of variability.

---

12 Wet oxidation of SO_2 by H_2O_2 and ozone was assumed in these calculations. HNO_3 was presumed to be captured nonreactively.

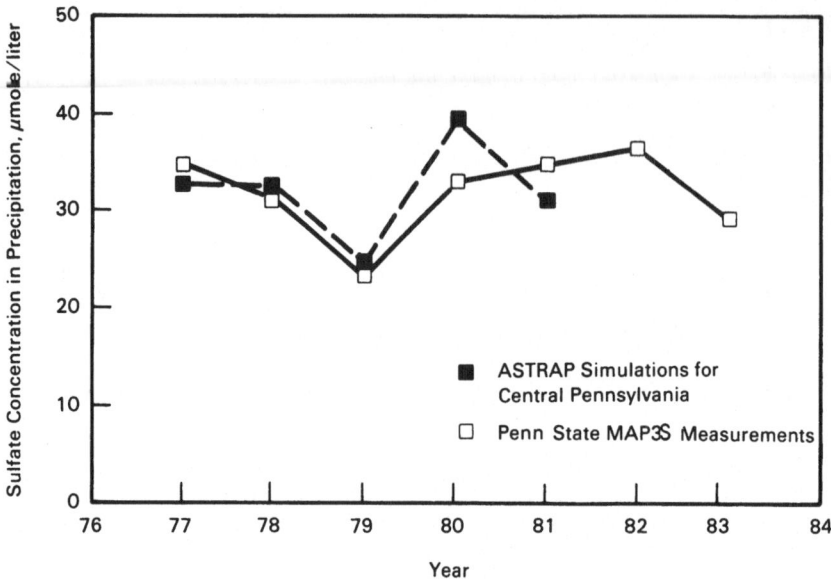

Fig. 22. Comparison of ASTRAP model predictions with annual averages of precipitation-borne sulfate concentrations observed at the Penn State MAP3S site [95]

Annually averaged sulfate-ion concentrations predicted by ASTRAP for central Pennsylvania are shown in Fig. 22, along with annually averaged Penn-State MAP3S data. The general variability exhibited by these results may be compared as well with longer-term observations shown in Figs. 5 and 6. Again it appears that the model has predicted variabilities that are comparable to observations, although the overall results have been "tuned" to a significant degree. These results again lend some confidence that the meteorological components of variability are generally well-characterized, although this says very little about the model's capability to describe source attribution accurately or to simulate the consequences of pronounced emission changes.

Spatial Variability Near Point Sources

Provided that background concentrations do not contribute to any significant degree, scavenging through point-source plumes presents a comparatively simple simulation problem, at least insofar as qualitative aspects of spatial variability of nonreactive systems are concerned. An example of such a simulation using the SMICK reactive plume-scavenging code (22) for SO_2 and sulfate ion in the vicinity of a coal-fired power plant is shown in Figs. 23 and 24. As can be noted from the first of these figures, fair agreement between predicted and observed SO_2 scavenging is achieved, both from the standpoints of absolute agreement of the values and qualitative depiction of the spatial variability.

Figure 24, which corresponds to sulfate scavenging from the plume, exhibits much worse agreement. Here the observed sulfate-ion concentrations (which have been corrected for background in the plot) maintain rather high central values and drop precipitously at the plume edges. The model calculations are based on

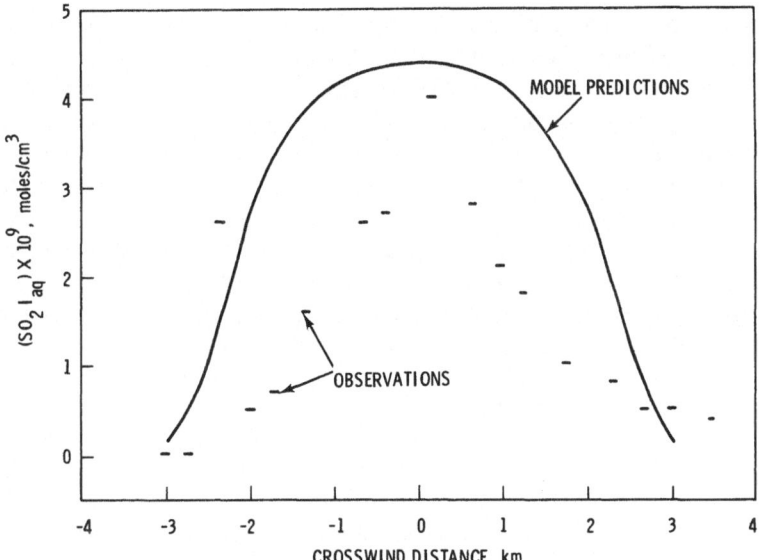

Fig. 23. Model-predicted and observed rainborne sulfate concentrations (background corrected) beneath the plume of a coal-fired power plant, at approximately 15 km from source. First-order SO_2 oxidation coefficients of 0.002 and $0.004\,s^{-1}$

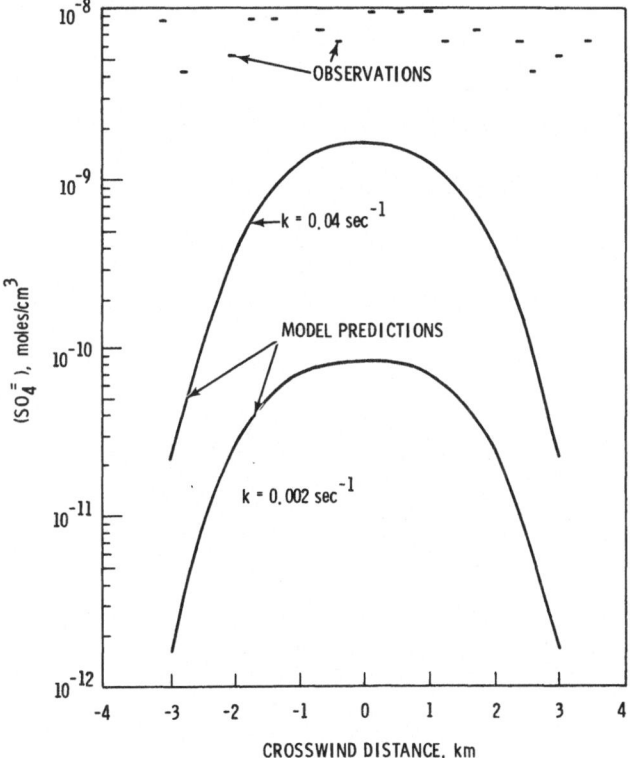

Fig. 24. Model-predicted and observed dissolved SO_2 concentrations in rain beneath the plume of a coal-fired power plant, at approximately 15 km from source

reactive scavenging of SO_2 to form sulfate, following a first-order rate process with the indicated coefficients. Obviously these computations are unsuccessful in matching both quantitative and qualitative aspects of the scavenging process.

The deficiency of the calculations for sulfate ion undoubtedly arose partially form the model's failure to account for scavenging of primary sulfate aerosol, which was emitted in small quantities concurrently with the SO_2. The contrast between the abrupt edges of the observed plume and those of its calculated counterparts, however, suggests that some sort of oxidant-limited mechanisms may have been in force. In any event, it appears that the simple simulation using assumed first-order oxidation chemistry has failed to give a good representation of the true chemical pathways involved in the scavenging process.

Spatial Variability on the Regional Scale

This final example of model-measurement comparison pertains to spatial distributions of averaged precipitation-borne sulfate concentrations predicted for the year 1980 by the ASTRAP model [95]. As noted previously, ASTRAP is a

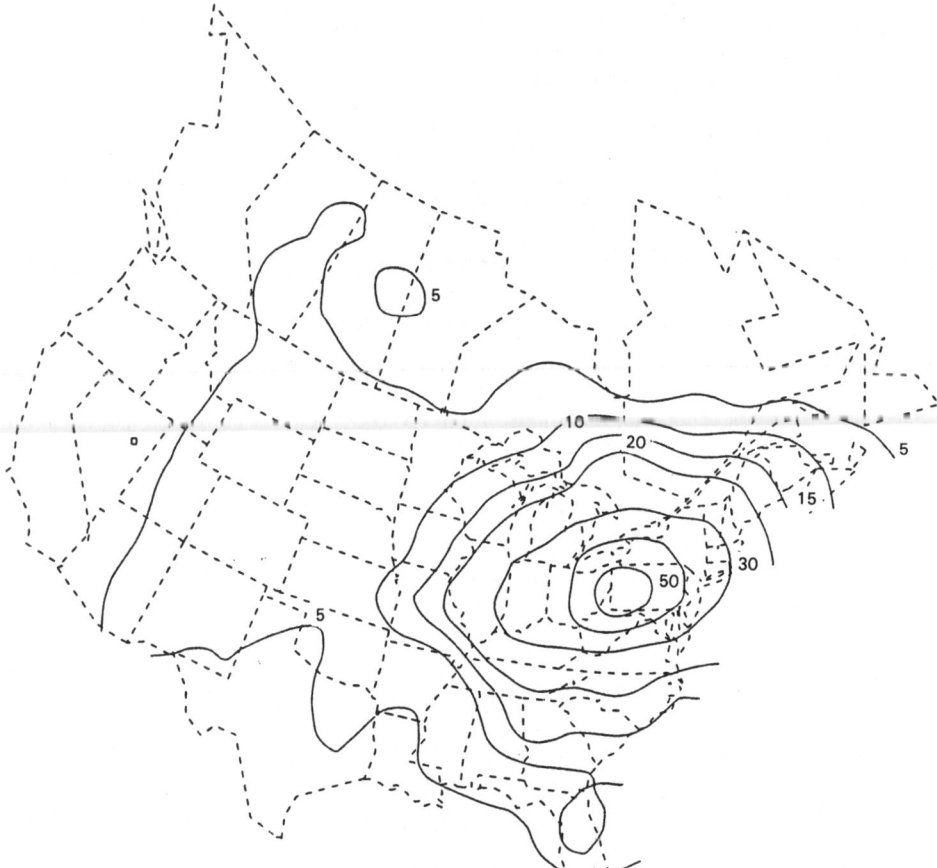

Fig. 25. ASTRAP model simulations of precipitation-borne sulfate concentrations for the year 1980 [95]. Isopleth values are in micromoles per liter

composite Lagrangian code which simulates the total source-receptor sequence of Fig. 1, and applies a modified scavenging-ratio method for wet-removal calculations. Input data for this simulation included the 1980 U.S./Canadian sulfur emission inventory and observed wind and precipitation fields.

The modeled results, shown in Fig. 25, can be compared directly with the observed data for 1980, which were presented in Fig. 9. In view of ASTRAP's relatively simplistic way of representing transformation and removal processes, the agreement between these two figures is rather impressive. Although this undoubtedly results from the model's fitted parameters to some extent, the fact that the spatial variations are matched quite well by the computed results suggest that the general characterization of transport and removal has been largely successful. This apparent success in pattern matching also depends on the fact that a large amount of averaging has occurred, and many of the previously noted variability components have been averaged out of the characterization. It should be noted as well that spatial-variability matching skill does not necessarily imply that the model possesses similar skill in source-attribution and rollback analyses. These are generally considered to be much more difficult features, from the standpoints of both simulation and evaluation against measured data.

Conclusion

This chapter has addressed the mathematical characterization of precipitation scavenging and precipitation chemistry, using a visual approach which focuses on spatial and temporal variability. In the first section both temporal and spatial variability were examined from a qualitative standpoint, using precipitation-chemistry observations obtained over telescoping scales of time and distance. Physical sources of these variabilities were examined in a qualitative sense, with the observation that various contributions to variability grow and diminish in importance as time and distance scales increase.

Subsequent sections have examined the mathematical basis for precipitation-scavenging models, and have demonstrated how the various components of variability appear in these formulations. This discussion begins by considering basic material, energy, and momentum balances, and progresses to the formulation of the models and the solution of their associated equations. Typical end-products of these solutions emerge as computed pollutant concentration fields and deposition fluxes. The discussion is rooted strongly on the procedure of "grooming" generalized property-balance equations to formulate the special forms characterizing the desired model. The final section provides some guidelines for selecting a pre-existing precipitation scavenging model to use in a specific situation. Brief examples of model-measurement comparisons are presented with the goal of providing the reader with an initial idea of model proficiency in this regard.

This chapter has been prepared with the intent of providing the reader with a good appreciation of precipitation scavenging, the general mathematical techniques used for its characterization, and the proficiency of today's models in generating realistic estimates of pollutant concentrations and delivery fluxes.

Upon mastering this material the reader may consider himself to have gained a fair proficiency in understanding causal relationships, evaluating the relative merits of existing models, and pursuing the current literature in the course of developing more advanced models for specific applications.

Acknowledgement

Preparation of this document was supported partly by the U.S. Department of Energy and partly by the U.S. Environmental Protection Agency. I express my sincere appreciation to these organizations for their assistance.

Although the research described in this article has been funded wholly or in part by the United States Environmental Protection Agency through a Related Services Agreement with the U.S. Department of Energy Contrac DE-AC06-76RLO 1830 to Pacific Northwest Laboratory, it has not been subjected to Agency review and therefore does not necessarily reflect the views of the Agency and no official endorsement should be inferred.

References

1. Adamowicz, R.F.: A Model for the Reversible Washout of Sulfur Dioxide, Ammonia, and Carbon Dioxide From a Polluted Atmosphere and the Production of Sulfate in Raindrops. Atm. Envi.*13*, 105 (1979)
2. Alp, E., Moran, M.D., Reid, N.W., McMillan, A.C.: Modelling of Long Range Transport and Acidic Deposition: Theoretical Evaluation of Models, Concord Scientific Corporation Report CSC J175.2 (1983)
3. Baker, M.B., Harrison, H., Vinelli, J., Erickson, K.B.: Simple Stochastic Models for the Sources and Sinks of Two Aerosol Types. Tellus *31*, 1 (1979)
4. Barrie, L.A.: An Improved Model of Reversible SO_2 Washout by Rain. Atm. Envi. *12*, 407 (1978)
5. Barrie, L.A., Hales, J.M.: The Spatial Distributions of Precipitation Acidity and Major Ion Wet Deposition in North America During 1980. Tellus *36* B, 353 (1984)
6. Bass, A.: Modeling Long-Range Transport and Diffusion. Proc. Second Conf. on Air Poll. Met. AMS/APCA, New Orleans 1980
7. Baum, S., Wong, P., Dolan, P. NUCROM: A Model of Rainout from Nuclear Clouds. SRI report to Defense Nuclear Agency. DNA001-73-C-0148 (1978)
8. Bhumralkar, C.M., Johnson, W.B., Mancusco, R.H., Thuiller, R.H., Wolf, D.E.: Interregional Exchanges of Airborne Sulfur Pollution and Deposition in Eastern North America. Proc. 2nd Conf. on Air Pollution Met. AMS/APCA New Orleans 1980
9. Bird, R.B., Stewart, W.E., Lightfoot, E.N.: Transport Phenomena. Wiley, New York 1960
10. Bolin, B., Persson, C.: Regional Dispersion and Deposition of Atmospheric Pollutants with Particular Application to Sulphur Pollution over Western Europe. Tellus *27*, 281 (1977)
11. Carmichael, G.R., Peters, L.K.: An Eulerian Transformation/Removal Model for SO_2 and Sulfate. Atm. Envi. *18*, 937 (1984)
12. Chamberlain, A.C.: Aspects of Travel and Deposition of Aerosols and Vapor Clouds. AERE Harwell, Report R1261 HMSO London 1953
13. Chameides, W.L., Davis, D.D.: The Free-Radical Chemistry of Cloud Droplets and its Impact Upon the Composition of Rain. J. Geophysics Research *87*, 4863 (1982)
14. Cogbill, C.V., Likens, G.E., Butler, T.A.: Uncertainties in Historical Aspects of Acid Precipitation: Getting it Straight. Atm. Env. *18*, 2261 (1984)
15. Dana, M.T., Easter, R.C., Thorp, J.M.: OSCAR Experiment High-Density Network Data Report, Event 4 April 21–23, 1981, Pacific Northwest Laboratory Report PNL-5152 Part 4 (1984)

16. Dana, M.T., Hales, J.M.: Statistical Aspects of the Washout of Polydisperse Aerosols. Atm. Env. *10*, 45 (1976)
17. Davies, T.D.: Rainborne SO_2, Precipitation pH and Airborne SO_2 Over Short Sampling Times Throughout Individual Events. Atm. Envi. *18*, 2499 (1984)
18. de Boer, J.H.: The Dynamical Character of Adsorption. Clarendon, Oxford 1953
19. de Pena, R.G., Carlson, T.N., Takacs, J.F., Holian, J.O.: Analysis of Precipitation Collected on a Sequential Basis. Atm. Env. *18*, 2665 (1984)
20. Dingle, A.N., Lee, Y.: An Analysis of In-Cloud Scavenging. J. Appl. Met. *12*, 1295 (1973)
21. Dingle, A.N., Lee, Y.: Terminal Fallspeeds of Raindrops. J. Appl. Met. *11*, 877 (1972)
22. Drewes, D.R., Hales, J.M.: SMICK – A Scavenging Model Incorporating Chemical Kinetics. Atm. Env. *16*, 1717 (1982)
23. Durham, J.L., Overton, J.H., Aneja, V.: Influence of Gaseous Nitric Acid on Sulfate Production and Acidity in Rain. Atm. Env. *15*, 1059 (1981)
24. Easter, R.C., Hales, J.M.: PLUVIUS: A Generalized One-Dimensional Model of Reactive Pollutant Behavior, Including Dry Deposition Pricipitation Formation, and Wet Removal, PNL-4046 ED2 Pacific Northwest Laboratory (1984). (Available through NTIS)
25. Easter, R.C., Hales, J.M.: Mechanistic Evaluation of Precipitation Scavenging Using a One-Dimensional Reactive Storm Model. Final Report to Electric Power Research Institute RP-2022-1 (1982)
26. Eliassen, A.: The OECD Study of Long-Range Transport of Air Pollutants. Atm. Envi. *12*, 479 (1978)
27. Eliassen, A., Hov, O., Isaksen, I.S.A., Saltbones, J., Stordal, F.: A Lagrangian Long-Range Transport Model with Atmospheric Boundary-Layer Chemistry J. Appl. Met. *21*, 1645 (1978)
28. Eliassen, A., Saltbones, J.: Modeling of Long-Range Transport of Sulphur over Europe: A Two-Year Model Run and Some Model Experiments. Atm. Envi. *17*, 1457 (1983)
29. Engelmann, R.J.: The Calculation of Precipitation Scavenging. Meteorology and Atomic Energy 1960, U.S. AEC, D. Slade, ed. (1968)
30. Engelmann, R.J.: Scavenging Precition Using Ratios of Air and Precipitation. J. Appl. Met. *10*, 493 (1971)
31. Fay, J.A., Rosenzwerg, J.J.: An Analytical Diffusion Model for Long-Distance Transport of Air Pollutants. Atm. Envi. *14*, 355 (1980)
32. Finkelstein, P.L.: The Spatial Analysis of Acid Precipitation Data. J. Clim. and Appl. Met. *23*, 52 (1984)
33. Fisher, B.E.A.: The Long-Range Transport of Sulphur Dioxide. Atm. Envi. *9*, 49 (1975)
34. Fitzgerald, J.W.: Effect of Aerosol Composition in Cloud Droplet Size Distribution: A Numerical Study. J. Atm. Sci. *31*, 1358 (1974)
35. Garner, F.H., Lane, J.J.: Mass Transfer to Drops Suspended in a Gas Stream. Trans. Inst. Chem. Engrs. *37*, 162 (1959)
36. Gibbs, A.G., Slinn, W.G.N.: Fluctuations in Trace Gas Concentrations in the Troposphere. J. Geophys. Res. *78*, 579 (1973)
37. Graedel, T.E., Goldberg, K.I.: Kinetic Studies of Raindrop Chemistry. J. Geophys. Res. *83*, 10865 (1983)
38. Granat, L.: Sulfate in Precipitation as Observed by the European Air Chemistry Network. Atm. Env. *12*, 413 (1978)
39. Greenfield, S.M.: Rain Scavenging of Radioactive Particulate Matter from the Atmosphere. J. Met. *14*, 115 (1957)
40. Hales, J.M.: A Generalized Multidimensional Model for Precipitation Scavenging and Atmospheric Chemistry. Atm. Envi. In Press (1985)
41. Hales, J.M., Fox, T., Powell, D.: An Air Pollution Model Incorporating Nonlinear Chemistry, Variable Trajectories, and Plume-Segment Diffusion. Battelle-Northwest Report to EPA, EPA-450/3-77-012 (1977)
42. Hales, J.M., Dana, M.T.: Regional-Scale Deposition of Sulfur Dioxide by Precipitation Scavenging. Atm. Envi. *13*, 1121 (1979)
43. Hales, J.M., Horst, T.W., Dana, M.T.: MPADD – A Generalized Code for Reactive Scavenging and Deposition Calculations Final Report Submitted to the Environmetal Protection Agency DW930197-01-1 (1983)

44. Hales, J.M.: Fundamentals of the Theory of Gas Scavenging by Rain. Atm. Env. 6, 635 (1972)
45. Hales, J.M., Dana, M.T.: Precipitation Scavenging of Urban Pollutants by Convective Storm Systems. J. Appl. Met. 18, 294 (1979)
46. Hales, J.M.: Precipitation Scavenging Processes. In The Acidic Deposition Phenomena and its Effects. A.P. Altschuller and R.A. Linthurst, eds., U.S. EPA 600/8-83-016AF (1984)
47. Hales, J.M.: Mechanistic Analysis of Precipitation Scavenging Using a One-Dimensional, Time-Variant Model. Atm. Env. 16, 1775 (1982)
48. Hales, J.M., Wolf, M.A., Dana, M.T.: A Linear Model for Predicting the Washout of Pollutant Gases from Industrial Plumes. AICHE Journal 19, 292 (1973)
49. Hane, C.E.: Scavenging of Urban Pollutants by Thunderstorm Rainfall: Numerical Experimentation. J. Appl. Met. 17, 699 (1978)
50. Hansen, D.A., Hidy, G.M.: Review Questions Regarding Rain Acidity Data. Atm. Env. 16, 2107 (1982)
51. Harrison, L.: The Segregation of Aerosols by Cloud-Nucleating Activity. Part II: Observation of an Urban Aerosol. J. Clim. & Appl. Met. 24, 4, 312–321 (1985)
52. Heffter, J.L.: Air Resources Laboratory Atmospheric Transport and Dispersion Model (ARL-ATAD). NOAA Technical Memorandum ER ARL-81 (1980)
53. Heffter, J.L.: Branching Atmospheric Trajectory (BAT) Model. NOAA Technical Memorandum ERL ARL-121 (1983)
54. Heikes, B.G., Thompson, A.M.: Effects of Heterogeneous Processes on NO_3, HONO, and HNO_3 Chemistry in the Troposphere. J. Geophys. Res. 88, 10, 883 (1983)
55. Hegg, D.A., Rutledge, S.A., Hobbs, P.V.: A Numerical Model for Sulfur Chemistry in Warm-Frontal Rainbands. J. Geophys. Res. 89, 7133 (1984)
56. Henmi, J.: Long-Range Transport Model of SO_2 and Sulfate and its Application to the Eastern United States. J. Geophys. Res. 85, 4436 (1980)
57. Hidy, G.: Removal of Gaseous and Particulate Pollutants. In Chemistry of the Atmosphere, S.I. Rasoule, ed. Plenum, NY (1973)
58. Hill, F.B., Adamowicz, R.F.: A Model for Rain Composition and the Washout of Sulfur Dioxide. Atm. Envi. 11, 912 (1979)
59. Hoffman, M.R., Calvert, J.G.: Chemical Transformation Modules for Eulerian Acid Deposition Models: Aqueous-Phase Chemistry. Report to U.S. EPA Interagency Agreement DW 930237 (1985)
60. Jensen, J.A., Charlson, R.J.: On the Efficiency of Nucleation Scavenging. Tellus 360, 36 B, 367 (1984)
61. Johnson, D.B.: The Role of Coalescence Nuclei in Warm Rain Initiation. Ph. D. Thesis, University of Chicago
62. Junge, C.E.: Air Chemistry and Radioactivity. Academic Press, New York 1963
63. Kleinman, L.J., Carney, J.G., Meyers, R.E.: Time Dependence on Average Regional Sulfur Oxide Concentrations Proc. 2nd Conf. on Air Poll. Met. AMS/APCA, New Orleans, 1980
64. Klett, J.: Precipitation Scavenging in Rainout Assessment: The ACRA System and Summaries of Simulation Results. LASL Report to ERDA, LA 6763, 1977
65. Knutson, E.O., Stockham, J.D.: Aerosol Scavenging by Snow: Comparison of Single-Flake and Entire Snowflake Results. In Precipitation Scavenging, ERDA Symposium series CONF 741003 (1977)
66. Kreitzburg, C.W., Leach, M.J.: Diagnosis and Prediction of Tropospheric Trajectories. Proc. 85th National Meeting, AIChE, Philadelphia, PA, 1978
67. Lange, R., Knox, J.B.: Adaptation of a Three-Dimensional Atmospheric Transport-Diffusion Model to Rainout Assessments. Precipitation Scavenging. R.S. Semonin and R.W. Bendle, eds. ERDA Symposium series CONF 741003 (1977)
68. Lavery, T.L.: Development and Validation of a Regional Model to Simulate Atmospheric Concentrations of SO_2 and Sulfate. Proc. 2nd Joint Conf. on Air Pollution Met., New Orleans, LA, 1980
69. Lee, H.N.: An Alternate Pseudospectial model for Pollutant Transport, Diffusion, and Disposition in the Atmosphere. Atm. Envi. 15, 1017 (1981)
70. Levine, S.Z., Schwartz, S.E.: In and Below-Cloud Scavenging of Nitric Acid Vapor. Atmos. Envi. 16, 1725 (1982)
71. Likens, G.E., Bormann, F.H., Pierce, R.S., Eaton, J.S., Munn, R.E.: Long-Term Trends in Precipitation Chemistry at Hubbard Brook, New Hampshire. Atm. Envi. 18, 2641 (1984)

72. Liljestrand, H.M., Morgan, J.J.: Error Analysis Applied to Indirect Methods for Precipitation Activity. Tellus *31*, 421 (1979)
73. Liu, M.K., Durran, D.: The Development of a Regional Air Pollution Model and its Application to the Northern Great Plains. EPA Report EPA-908/1-77-001 (1977)
74. Mason, B.J.: The Physics of Clouds. Clarendon, Oxford 1971
75. MAP3S/RAINE Research Community. The MAP3S/RAINE Precipitation Chemistry Network: Statistical Overview for the Period 1976–1980. Atmospheric Environment *16*, 1603 (1982)
76. McNaughton, D., Powell, D., Berkowitz, C.: A User's Guide to RAPT. MAP3S/RAINE Report PNL-3390 (1981)
77. Mellor, G.L., Yamada, T.: A Hierarchy of Turbulence-Closure Models for Planetary Boundary Layers. J. Atmos. Sci. *31*, 1791 (1974)
78. Molenkamp, C.R.: A Scavenging Model for Stratified Precipitation in Precipitation Scavenging, Dry Deposition and Resuspension. Pruppacher, Semonin, and Slinn, eds. Elsevier, New York 1983
79. OECD: Long-Range Transport of Air Pollutants. Organization for Economic Cooperation and Development. Paris (1977)
80. Overton, J.H., Aneja, V., Durham, J.L.: Production of Sulfate in Rain and Raindrops in Polluted Atmospheres. Atm. Envi. *13*, 355 (1979)
81. Patterson, D.E., Husar, R.E., Wilson, W.E., Smith, L.F.: Monte Carlo Simulation of Daily Regional Sulfate Distribution. J. Appl. Met. *20*, 404 (1981)
82. Peters, L.K.: Gases and Their Precipitation Scavenging in the Marine Atmosphere. In Air-Sea Exchange of Gases and Particles, P.S. Liss and W.G.N. Slinn, eds. Reidel, Dordrecht 1983
83. Prahm, L.P., Christensen, O.: Longe-Range Transmission of Pollutants Simulated by a Two-Dimensional Pseudospectral Dispersion Model. J. Appl. Met. *16*, 896 (1977)
84. Pruppacher, H.R., Klett, J.D.: Microphysics of Precipitation. Reidel, Dordrecht 1980
85. Radke, L.F., Hobbs, P.V., Eltgroth, M.W.: Precipitation Scavenging of Aerosol Particles by Precipitation. J. Appl. Met. *19*, 715 (1980)
86. Rodhe, H., Granat, L.: An Evaluation of Sulfate in European Precipitation, 1955–1982. Atm. Envi. *18*, 2727 (1984)
87. Rodhe, H., Grandell, J.: On the Removal Time of Aerosol Particles from the Atmosphere by Precipitation Scavenging. Tellus *24*, 442 (1972)
88. Rodhe, H., Grandell, J.: Estimates of Characteristics Times for Precipitation Scavenging. J. Atm. Sci. *38*, 370 (1981)
89. Sampson, P.J.: Trajectory Analysis of Summertime Sulfate Concentrations in the Northeastern United States. J. Appl. Met. *19*, 1382 (1980)
90. Scott, B.C.: Parameterization of Sulfate Wet Deposition. J. Appl. Met. *17*, 1375 (1978)
91. Schwartz, S.E.: Mass-Transport Considerations Pertinent to Aqueous-Phase Reactions of Gases in Liquid-Water Clouds. NATO Advanced Study Institute on Chemistry of Multiphase Atmospheric Systems. Corfu, Greece, 1983
92. Scriven, R.A., Fisher, B.E.A.: The Long-Range Transport of Airborne Material and its Removal by Deposition and Washout. Atm. Envi. *9*, 49 (1975)
93. Seigneur, C., Saxena, P.: A Study of Atmospheric Acid Formation in Different Environments. Atm. Envi. *18*, 2109 (1984)
94. Shannon, J.: A Regional Model of Long-Term Average Sulfur Atmospheric Pollution, Surface Removal, and Net Horizontal Flux. Atm. Envi. *5*, 689 (1981)
95. Shannon, J.D., Streets, D.G.: Climatological Variability in the Evaluation of Cost Effectiveness of Emission Control Strategies to Reduce Acid Deposition. 15th NATO/CCMS International Technical Meeting on Air Pollution Modeling and its Applications. St. Louis (1985)
96. Sherwood, T.K., Pigford, R.L.: Absorption and Extraction McGraw-Hill, NY 1952
97. Sherwood, T.K., Pigford, R.L., Wilke, C.R.: Mass Transfer McGraw-Hill, NY 1975
98. Slade, D.H.: Meteorology and Atomic Energy. USAEC TID-24190 (1968)
99. Slinn, W.G.N.: Air-to-Sea Transfer of Particles. In Air-Sea Exchange of Gases and Particles, P.S. Liss and W.G.N. Slinn, eds. Reidel, Dordrecht 1983

100. Slinn, W.G.N.: Some Approximations for the Wet and Dry Removal of Particles and Gases from the Atmosphere. Water, Air, and Soil Poll. *7*, 513 (1977)
101. Slinn, W.G.N.: The Redistribution of A Gas Plume Caused by Reversible Washout. Atm. Envi. *8*, 233 (1974)
102. Slinn, W.G.N.: Precipitation Scavenging. In Atmospheric Science and Power Production, D. Randerson, ed. U.S. Department of Energy (NTIS DOE/TIC-2760) (1984)
103. Storebo, P.B., Dingle, A.N.: Removal of Pollution by Rain in a Shallow Air Flow. J. Atm. Sci. *31*, 533 (1974)
104. Voldner, E.C., Olson, K., Oikawa, K., Loiselle, M.: Comparison Between Measured and Computed Concentrations of Sulfur Compounds in Eastern North America. J. Geophys. Res. *86*, 5339 (1981)
105. Walcek, C.J., Pruppacher, H.R.: On the Scavenging of SO_2 by Cloud and Raindrops. J. Atm. Chem. *1*, 269 (1984)
106. Wang, P.K., Pruppacher, H.R.: An Experimental Determination of the Efficiency with which Aerosol Particles are Collected by Water Drops in Subsaturated Air. J. Atm. Sci. *34*, 1664 (1977)
107. Wangen, L.E., Williams, M.D.: Elemental Deposition Downwind of a Coal-Fired Power Plant. Water, Air, and Soil Pollution *10*, 33 (1978)
108. Watson, C.W., Barr, S., Allenson, R.E.: Rainout Assessment: The ACRA System and Summaries of Simulation Results. Los Alamos Scientific Laboratory Report 1977
109. Wilkening, K.E., Ragland, K.W.: User's Guide for the University of Wisconsin Atmospheric Sulfur Computer Model (UWATM-SOX) Report to EPA/Duluth Research Laboratory (1980)

Nomenclature

Variables	Definitions
a	radius of aerosol particle
c	pollutant concentration (moles per unit volume of total space)
$\hat{c}$	pollutant concentration in a single hydrometeor (moles per unit volume of water)
$\hat{C}$	mixed-average concentration of pollutant in an ensemble of hydrometeors existing in a volume of space at any given time (moles per unit volume of water)
$\hat{\tilde{C}}$	mixed-average concentration of pollutant in an ensemble of hydrometeors which have fallen through a given horizontal area during a specified period of time (moles per unit volume of water)
C_p, C_v	specific heats at constant pressure and volume (molar basis)
D	molecular (or Brownian) diffusivity
D_e	equivalent diameter of snowflake
E	efficiency for capture of pollutant by a single falling hydrometeor
e	individual capture efficiency for a specified physical mechanism
f	probability-density function
F	general functional operator
F_A	interfacial flux of pollutant molecules
$\underset{\sim}{F}_v$	viscous-force vector in momentum equation
$\underset{\sim}{g}$	gravitational acceleration vector
g	vertical component of gravitational acceleration vector
g_c	gravitational conversion factor
h	plume source height
$\underset{\sim}{i}, \underset{\sim}{j}, \underset{\sim}{k}$	Cartesian unit vectors
J_{rain}	rainfall rate (or rainfall volume flux)
k	reaction-rate coefficient; also individual mass-transfer coefficient
$\underset{\sim}{K}$	eddy diffusivity tensor

Variables	Definitions
K	component of eddy diffusivity tensor
K_y	overall mass-transfer coefficient, based on gaseous-phase driving force
m	mass of aerosol particles per unit volume of space
M	molecular weight of air
N_T	number of physical elements (drops, aerosol particles, ...) per unit volume of space
p	atmospheric pressure
Q	pollutant source strength
r	molar mixing ratio
r^*	physicochemical reaction rate
R	gas-law constant
$\tilde{R}$	radiant heating rate
Re	Reynolds number
S	Stokes number
Sc	Schmidt number
Sh	Sherwood number
t	time
T	temperature
U, V, W	velocity components in x, y, and z directions
$\underset{\sim}{v}$	velocity vector
v_d, v_w	equivalent dry and wet deposition velocities
$\hat{V}$	ensemble-average raindrop velocity
V_{rain}	volume of hydrometeors per volume of space
w	sedimentation velocity of hydrometeors (positive upward)
w^*	interphase mass-transport rate of pollutant
x, y, z	Cartesian scalar distances
X, Y, Z	domain limits in x, y, and z directions
α	solubility (gas-liquid partition coefficient)
γ_d	dry-adiabatic lapse rate
Λ	scavenging coefficient
Λ_r	reversible scavenging coefficient
ε	storm extraction efficiency
λ	latent heat of phase transformation
ξ	scavenging ratio
ϱ	air density
σ_y, σ_z	plume-spread parameters
τ_w, τ_d	characteristic residence times for pollutants in atmosphere, associated with wet and dry deposition

Subscripts

$A, B, ...$	pollutant species
c	condensation
d	vapor deposition to ice; also "dry"
f	freezing of liquid water; also "forward"
i	"interfacial"; also vector component
m	physical medium
r	"reverse"
w	"wet"
x	direction indicator, also "aqueous phase"
y	direction indicator, also "gaseous phase"
z	direction indicator

Subject Index